Textbook of Phycology and Bryology

The Author

Dr. K. Sivakumar, M.Sc. (Botany). M.Sc. (Biolnform.), M.Phil. Ph.D. (Botany) is working as a Professor in the Department of Botany, Annamalai University, having 23 years teaching and 30 years research experience in the fields of electron microscopy, algal taxonomy, ecology, biotechnology and bioinformatics. He completed his Post-Doctoral work in South Korea and has published more than 103 research articles in international and national journals. Dr. SIVAKUMAR has been invited by Sri Lanka, Japan, Taiwan, Korea and Australia for his proficiency in botanical research. He is also a member in several professional and academic bodies.

Dr. Sivakumar has successfully guided 22 M.Phil, and 17 Ph.D. scholars and is currently guiding 01 Ph.D. students. He has completed four major research project sponsored by the University Grants Commission, Department of Science and Technology and Indian Council of Medical Research (ICMR), India. Registered one copyright for Lord Ganesh by Marine Algae through: IPR-Ministry of Higher Education, Govt, of India, New Delhi.

He has organized one National Conference on "Future Perspectives on Botanical Research", sponsored by UGC and DST, during 2011 and published an abstract manual and one National workshop on "Biological specimen preparation for SEM studies", sponsored by UGC, during 2014 and published a workshop manual. He also published a book entitled "Ultrastructural studies on *Hypnea lamourx* (*Gigartinales, Rhodophyceae*)" in M/s Bishen Singh Mahendra Pal Singh Publication, Dehradun, 2015, "Algae - A Practical Approach", 2016, MJP Publishers, Chennai and "Biodiversity and Biology of Fresh Water Algae", 2019, Saras Publication, Nagercoil.

Textbook of Phycology and Bryology

by

Dr. K. Sivakumar

2025

Daya Publishing House®

A Division of

Astral International Pvt. Ltd.

New Delhi – 110 002

ISBN: 978-93-5919-526-1

Published by : **Daya Publishing House®**
A Division of
Astral International Pvt. Ltd.
– ISO 9001:2015 Certified Company –
4736/23, Ansari Road, Darya Ganj
New Delhi-110 002
Ph. 011-43549197, 23278134
E-mail: info@astralint.com
Website: www.astralint.com

Printed at : **Replika Press**

PREFACE

The chief aim of writing this book is to meet the requirements of the botany students of degree classes of the various Indian Universities. It is the outcome of considerable experience of the author in teaching the subject to the college classes. It is proposed to bring out the book in five units.

The first unit deals briefly with the important system of classification of the plant kingdom and a general account of the thallophytes. The succeeding units are concerned with the detailed descriptions of the life histories of the important representative of each class of algae. Suitable and properly labelled diagrams have been given to illustrate the various stages in the life histories of the representative types. A list of diagnostic features of important types and graphic representation of their life histories have been given at the end of each chapter to enable the readers to have a clear cut idea of the type described. With the same object in view a self-learning materials has been added at the end of each unit. The questions will help the readers to grasp the subject matter fully. The fouth and fifty unit include a general account of the various aspects of the life cycles of Bryophytes. The subject matter has been arranged in what seems to the author a logical order. Care has also been taken to bring the subject matter up-to-date keeping in view the latest researches on the subject. Every attempt has been made to write in a simple and lucid style. Difficult phrases and lengthy sentences have been, so far as possible, avoided.

The author lays no claim to originality in the preparation of the book. It is a compilation work done in a manner in which it would prove most useful to the reader. All suggestions to improve the book will be thankfully accepted.

I am thankful to my publishers for providing me all the facilities in the publication to bring out the Text Book of Phycology and Bryology.

Dr. K. Sivakumar

CONTENTS

Unit 1

Algae

Introduction

1. The algae are chlorophyll-bearing autotrophic organisms with a thallus-like plant body.
2. The thallus shows little differentiation of true tissues.
3. Even the complex thalli lack vascular tissue and epidermis with stomata
4. The sex organs are one-celled, when multicellular, each cell is fertile and there is no jacket of sterile cells.
5. There is no embryo formation after gametic union.
6. Both the generations when represented in the life cycle are independent. There are no algae with a sporophyte parasitic on the gametophyte plant.
7. Excepting a few all the algae are aquatic.
8. Under favourable conditions the gametophyte multiplies repeatedly by means of asexual spores called the mitospores.

Algal characteristics basic to primary classification. The primary classification of algae is based on certain morphological and physiological features. The chief among these are: (*a*) pigment constitution of the cell, (*b*) chemical nature of stored food materials, (*c*) kind, number, point of insertion and relative length of the flagella on the motile cell, (*d*) chemical composition of cell wall and (*e*) presence or absence of a definitely organised nucleus in the cell or any other significant detail of cell structure. The importance of these features in the primary classification of algae has been discussed in the preceeding pages. The details of vegetative structure and reproduction are useful for algal classification only at the level of families, genera and species.

General Characteristics

The term 'algae' (singular alga) is a Latin word that means seaweed. The science that deals with the study of algae is called *algology.* The Greek word for algae is *phykos* (literally, seaweed), and, therefore, the study of algae is also called phycology. For a pretty long time, algae were popularly known by the name *fucus.* as Romans pronounced phykos as *fucus.* Chinese named them *tsao* and Hawaiians *limu.* Laymen have named algae variously as pond scums, frog spittle, water mosses, stoneworts and seaweeds.

Algae comprise a heterogeneous assemblage of relatively simple plants, including about 30,000 species. HC Bold (1967) reported that there are only 21,000 species in algae. They evolved during Ordovician of Palaeozoic era, the period popularly called 'age of seaweeds'. As evidenced from molecular genetics, the land plants, such as bryophytes and tracheophytes, may have arisen independently from marine algae related to the present-day Chlorophyta (green algae) and Charophyta.

Some phycologists have defined algae as follows:

F E Fritsch (1935) "Unless purely artificial limits are drawn, the designation alga must include all holophytic organisms that fail to reach the higher level of differentiation characteristic of the archegoniateplants".

G M Smith (1955) "Simple plants with an autotrophic mode of nutrition".

V J Chapman (1962) "These plants which are among the simplest in the plant kingdom, belong to the group known as algae".

G M Prescott (1969) "The algae are those chlorophyll bearing organisms which are thalloid".

RE Lee (2008) "The algae are thallophytes that have chlorophyll as their primary photosynthetic pigment and lack a sterile covering of cells around the reproductive cells".

On the basis of their characteristic features this author defines algae as under: "A heterogeneous group of first oxygen evolving aquatic autotrophic thallophytes which lack sterile jacket around the reproductive cells".

Algae incorporate several unicellular divisions, a prokaryotic division Cyanophyta (Cyanobacteria-blue green algae); three main multicellular divisions: Rhodophyta (red algae), Chlorophyta (green algae) and Charophyta, and a class Phaeophyceae (brown algae). Unicellular divisions include planktonic diatoms (appearing colourless) and dinoflagellates that are responsible for most photosynthesis in the oceans, and therefore of the world. Most algae are aquatic (both freshwater and marine).

Important Classification of Algae

The committee on the International Code of Botanical Nomenclature has recommended certain suffixes for use in the classification of Algae. These are phyta for *division,* phyceae for *class,* phycideae for *subclass,* ales for *order,* inales for *sub-order,* aceae for *family,* oideae for *sub-family,* Greek name for *genus* and Latin name for a *species.*

Classification Proposed by Fritsch

The first most comprehensive and authorative classification of algae was given by F.E. Fritsch (1935,48) in his book *'The Structure; and Reproduction of the Algae'.* His classification was based on such criteria as pigmentation, types of flagella, assimilatory products, thallus structure and methods of reproduction. He divided algae into the following 11 classes:

1. Class: Chlorophyceae (Isokontae, green algae)

The main pigments are essentially identical to those found in higher plants. They are chlorophyll *a* and *b,* carotenoid and xanthophylls. The chloroplast usually has pyrenoids. The food reserve is starch, which frequently aggregates around the pyrenoids. The cell wall is composed of cellulose, and the cells are typically eukaryotic. Flagella, if present, are of equal length and are usually inserted at the anterior end. Reproduction takes place by vegetative, asexual and sexual (iso-, aniso-, and oogamous) methods.

The class Chlorophyceae has been divided into nine orders:

(i) Volvocales (*e.g., Volvox*),

(ii) Chlorococcales (*e.g., Chlorella*),

(iii) Ulotrichales (*e.g., Ulothrix*),

(iv) Cladophorales (*e.g., Cladophora*),

(v) Chaetophorales (*e.g., Fritschiella*),

(vi) Oedogoniales (*e.g., Oedogonium*),

(vii) Conjugates (*e.g., Zygnema* and *Spirogyra*),

(viii) Siphonales (*e.g., Vaucheria* and *Botrydium*),

(ix) Charales (*e.g., Chara* and *Nitella*)

2. Class: Xanthophyceae (Heterokontae, yellow green algae)

The main pigments are chlorophyll *a* and *b,* β-carotene and xanthophylls. The plastids are without pyrenoids and the reserve food material is oil. The cell wall is mainly composed of pectic substances with little cellulose. The cell is typical eukaryotic with two anteriorly inserted unequal flagella. The shorter flagellum is of whiplash type and the longer one of tinsel type. Reproduction takes place by vegetative, asexual and sexual (isogamous) methods.

This class has been divided into four orders: (i) Heterochloridales (*e.g., Heterochloris, Chlora-moeba*), (ii) Heterococcales (*e.g., Myxochloris, Halosphaera*), (iii) Heterotrichales (*e.g., Tribo-nema, Microspore*), and (iv) Heterosiphonales (*e.g., Botrydium*).

3. Class: Chrysophyceae

Phycochrysin is the dominant pigment which imparts brown or orange colour to these algae. The chromatophores have naked pyrenoid-like bodies. The food reserve is chrysolaminarin and leucosin. The cell wall is silicified or calcified and non-cellulosic. The motile cells have two anteriorly inserted equal or unequal flagella. Sexual reproduction is very rare; when present then it is isogamous.

The class includes three orders: (i) Chrysomoeadales (*e.g., Chrysococcus, Chryso-dendron, Chromulina*), (ii) Chrysosphaerales (*e.g., Chrysosphaera, Echinochrysis*), (iii) Chryso-tricfiales (*e.g., Nematochrysis, Chrysoclonium*).

4. Class: Bacillariophyceae (Diatom, yellow or golden-brown algae)

The members of this class are characterized by the dominance of golden-brown pigments, *viz.*, fucoxanthin, diatoxanthin and diadinoxanthin. The chromatophores have pyrenoids and the photosynthetic products are fat and volutin. The cell wall is pectic and silicified and variously ornamented. Usually it consists of two halves which are radially or bilaterally symmetrical. The motile cells usually have a single flagellum. The sexual reproduction takes place by fusion and by the formation of gametes or auxospores.

The class includes two orders: (i) Centrales (*e.g., Cyclotella, Chaetoceras*), and (ii) Pennales (*e.g., Nitschia, Navicula, Pinnularia, Denticula, Epithemia*).

5. Class: Cryptophyceae

The main pigment is xanthophyll which imparts brown or red colour to these algae. Pyrenoid-like bodies are present but they are often independent of chromatophores. The photosynthetic products are starch and or oil. The motile cells are dorsiventral with two anteriorly inserted unequal flagella. A complex vacuolar system is present in the cell. Sexual reproduction is rare and when present, it is of isogamous type. The class includes two orders: (i) Cryptomonadales (*e.g., Cryptomonas, Rhodomonas, Cyanomonas*), and (ii) Cryptococcaales (*e.g., Tetragonidiwri*).

6. Class: Dinophyceae (Peridinieae)

These algae contain many discoid chromatophores and the main pigment is xanthophyll which imparts brown or red colour. The food reserves are starch and fats. The cell wall is cellulosic. Most of the members of this class are motile unicells with two flagella. Many species are colourless saprophytes or show holozoic mode of nutrition. Sexual reproduction is rare and is of isogamous type, if present. They also multiply by cell division.

In this class six orders are included: (i) Desmomonadales (*e.g., Desmocapsa, Pleromonas, Desmomastix*), (ii) Thecatales (*e.g., Exuviaella, Prorocentrum*), (iii) Dinophysiales (*e.g., Dinophysis, Phalacroma*), (iv) Dinoflagellata (*e.g., Blastodinium, Ceratium, Heterocapsa*), (v) Dinococcales (*e.g., Dinastridium, Cystodinium, Dissodinium*), and (vi) Dieotrichales (*e.g., Dinothrix, Dinoclonium*).

7. Class: Chloromonadineae

The members of this class have numerous discoid chromatophores; they provide bright green tint due to the presence of excess of xanthophyll. Pyrenoids are lacking and the reserve food is fat and oil. The motile forms are with two almost equal flagella. Sexual reproduction is absent; multiplication takes place by longitudinal division of cells. This is a small class with only one order, Chloromonadales, comprising of highly specialized unicellular flagellates (*e.g., Trentonia, Vacuolaria*).

8. Class: Euglenineae

These are highly specialized unicellular green flagellates. The cell has usually many chromatophores. Pyrenoid-like bodies are present in some forms. The main pigment is chlorophyll, and the product of photosynthesis is a polysaccharide, paramylon. Most of the members have one or two flagella which arise from the base of a canal like invagination at the anterior end. These algae possess a complex vacuolar system. The multiplication is by cell division. Sexual reproduction is known only in few forms and is of isogamous type. This class includes 'three families: (i) Euglenaceae (*e.g., Amphitropis, Ascoglena, Euglena*); (ii) Astasiaceae (*e.g., Astasia, Distigma, Menoidium*), and (iii) Peranemaceae (*e.g., Anisonema, Entosiphon, Petalomonas*).

9. Class: Phaeophyceae (brown algae)

The chromatophores, besides other pigments, contain fucoxanthin. Some lower forms have naked pyrenoid- like bodies. The food reserves are in the form of alcohol (mannitol), polysaccharides (laminarin) and fats. The cell wall is mainly composed of cellulose with alginic and fucinic acids. The motile reproductive cells have two lateral or sub-apical flagella, one directed forward and the other backward. The sexual reproduction ranges from isogamous to oogamous type.

The class comprises of nine orders: (i) Ectocarpales (*e.g., Ectocarpus, Halothrix, Punctaria*), (ii) Tilopteridales (*e.g., Tilopteris*), (iii) Cutleriaies (*e.g., Cutleria*), (iv) Sporochnales (*e.g., Sporochnus, Carpomitra*), (v) Desmarestiales (*e.g., Desmarestia, Arthrocladia*), (vi) Laminariales (*e.g., Laminaria, Lessonia, Macrocystis*), (vii) Sphacelariales (*e.g., Sphacelaria, Halopteris*), (viii) Dictyotales (*e.g., Dictyota*), and (ix) Fucales (*e.g., Fucus, Sargassum, Durvillea*).

Some Salient Features of Major Classes of Algae

Sl.No.	*Characters*	*Cyanophyceae*	*Chlorophyceae*	*Bacillariophyceae*	*Phaeophyceae*	*Rhodophyceae*
1.	**Habitat**	Fresh water, marine or terrestrial.	Fresh water, marine or terrestrial.	Fresh water, marine or terrestrial.	Only four genera are fresh water, rest marine.	Mostly marine; a few are terrestrial.
2.	**Thallus habit**	Unicellular, colonial or filamentous.	Unicellular, colonial, coenobial, filamentous, thalloid or siphonaceous.	Unicellular or colonial.	Filamentous, branched, usually heterotrichous.	Unicellular, filamentous, pseudoparen-chymatous.
3.	**Cell structure**	Prokaryotic, membrane bound nucleus and other cell organelles like plastids, mitochondria, *etc.*, are absent	Eukaryotic, well organised nucleus and other cell organelles present.	eukaryotic.	eukaryotic.	eukaryotic.
4.	**Cell wall**	Mucopeptides and muramic acid are two main constituents of the cell wall.	Cell wall is made up of cellulose and pectic material.	Cell wall is made up of silica and pectic material.	Outer mucilagenous layer of cell wall consists of alginic and fucinic acids and the inner layer is made up of cellulose.	Outer layer is pectic in nature whereas inner layer is cellulosic.
5.	**Flagella**	Non flagellate, motile cells are absent.	Motile cells with 2, 4 or more flagella; flagella anteriorly inserted, equal in length and whiplash type.	Usually 1, or sometimes 2, anteriorly inserted.	2, laterally inserted, of unequal length; one of them whiplash and another of tinsel type.	Non flagellate, motile cells are absent.

Sl.No.	*Characters*	*Cyanophyceae*	*Chlorophyceae*	*Bacillariophyceae*	*Phaeophyceae*	*Rhodophyceae*
6	**Dominant pigments**	Chlorophyll α, β-carotene, xanthophyll and c-phycocyanin.	Chlorophyll α and *b*, α- and β-carotene, xanthophyll.	Chlorophyll α, and c, β-carotene, xanthophyll (fucoxanthin).	Chlorophyll α, and c, β-carotene, xanthophyll (violaxanthin, fucoxanthin).	Chlorophyll α, and c, β-carotene (lutein), xanthophylls, r-phycoerythrin. cphycocyanin
7.	**Kinds of plastid**	No plastids	Chloroplast	Chromatophores	Chromatophores	Rodoplast or Chromatophore
8.	**Nucleus**	Nuclear membrane absent, DNA not associated with histones, therefore well-organized chromosomes are not found.	Well organized nucleus with chromosomes and nucleolus is present.	As in Chlorophyceae.	As in Chlorophyceae.	As in Chlorophyceae.
9.	**Eye spot**	Absent.	Attached with chloroplast but it has no relation with flagella.	Absent.	In male gametes, eye spot is present, attached to chromatophores. it is related with flagella.	Absent.
10.	**Pyrenoid**	Absent.	Single or compound, present either on the surface or embedded in chloroplast.	Simple or compound; one or more.	Single, stalked or projected.	Simple or compound; embedded, sometimes stalked.

Sl.No.	Characters	Cyanophyceae	Chlorophyceae	Bacillariophyceae	Phaeophyceae	Rhodophyceae
11.	**Reserve food material**	Special type of starch (cyanophycean starch) and protein (cyanophycin granules).	Starch and fats.	Oil, volutin and chrysose.	Laminarin, mannitol, sucrose.	Floridean starch.
12.	**Reprodu-ction**	Multiplication only through asexual methods; sexuality completely absent.	Reproduction by vegetative, asexual and sexual means; sexual reproduction isogamous, anisogamous or oogamous.	Reproduction by vegetative (cell division) and sexual means; sexual reproduction iso-, aniso-, or oogamous.	Reproduction by vegetative (fragmentation, adventitious branches), asexual (motile and aplanospores), sexual reproduction iso-, aniso-, or oogamous.	Reproduction by vegetative (cell division, fragmentation). asexual (monospores, neutral spores, tetraspores, carpospores) and sexual means; sexual reproduction complex and advanced oogamous.

10. Class: Rhodophyceae (red algae)

Most of these algae are marine with uniaxial or multiaxial thalli. The chromatophores, besides other pigments, contain r-phycoerythrin and r-phycocyanin which impart red colour to these algae. Pyrenoid-like bodies occur in the lower forms of this class. The food reserve is a solid polysaccharide, the floridean starch. The outer cell wall is pectic and the inner cellulosic. The reproductive cells are nonmotile; the sexual reproduction is of advanced oogamous type.

The class is divided into seven orders: (i) Bangiales (*e.g., Bangia, Porphyra, Porphyridium*), (ii) Nemalionales (*e.g., Batrachospermum, Lemanea, Nemalion*), (iii) Gelidiaies (*e.g., Gelidium*), (iv) Cryptonemiales (*e.g., Dumontia, Corallina, Gloeopeltis*), (v) Gigartinales (*e.g., Ahnfeltia, Chondrus, Gigartina, Gracilaria*), (vi) Rhodymeniales (*e.g., Champia, Rhodymenia, Gastroclonium*), and (vii) Ceramiales (*e.g., Ceramium, Polysiphonia, Lophosiphonia*).

11. Class: Myxophyceae (Cyanophyceae, blue-greee algae)

These algae are characterized by the presence of a very rudimentary nucleus and they do not have well organized chromatophores (*i.e.*, cell organization is prokaryotic). The chief pigments are chlorophyll a, β-carotene and c-phycocyanin. The cell wall is made up of mucopolymers. The food reserve is cyanophycean starch. Sexual reproduction is absent (recently genetic recombination has been reported); asexual reproduction is by hormogonia or akinetes.

This class comprises of five orders: (i) Chroococcales (e,g., *Chroococcus, Gloeocapsa, Microcystis*), (ii) Chamaesiphonales (*e.g., Chamaesiphon, Dermocarpa*), (iii) Pleurocapsales (*e.g., Pleurocapsa*), (iv) Nostocales (*e.g., Gloeotrichia, Nostoc, Oscillatoria, Rivularia, Scytonema, Spirulina*), and (v) Stigonematales (*e.g., Stigonema, Mastigocladium, Nostochopsis*).

Thallus Construction Range of Structures: Evolutionary Trends of Chlorophyceae

Algae are chlorophyll containing primitive plants both prokaryotic and eukaryotic, with wide range of thalli starting from unicellular to multicellular organistions. They are autotrophic in nature.

Fritsch, F.E. (1935) defined algae as the holophytic organisms that fail to reach the higher level of differentiation characteristic of the archgoniate plants.

He proposed a most comprehensive and authorative classification of algae in his book "The Structure and Reproduction of the Algae", based on the following characters:

- ☆ Pigmentation
- ☆ Metabolic or assimilating food products
- ☆ Types of Flagella

Classified algae into 11 classes:

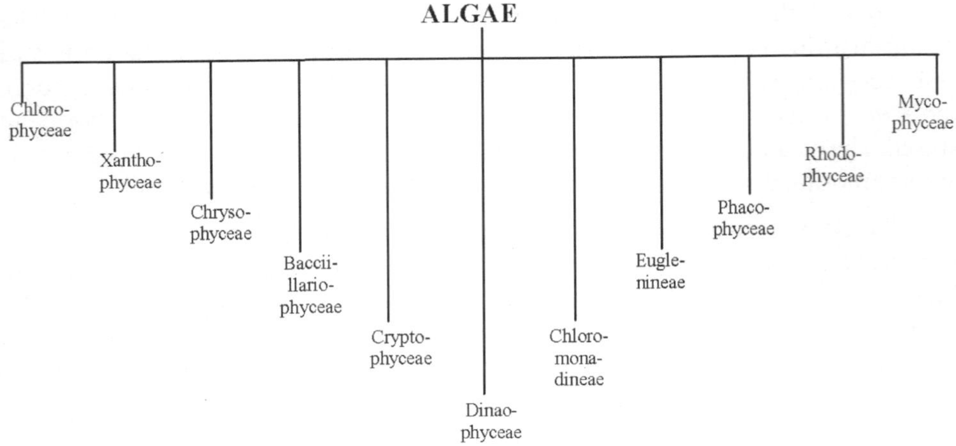

Chlorophyceae: Green Algae

- ☆ Chlorophyceae (chloros : green; phyceae : algal organization) is commonly known as green algae.
- ☆ Grow mostly in fresh water, a few in brackish and saline water, few are terrestrial.
- ☆ Show wide range of variations in their thallus structures like unicellular motile and non-motile, coenobium, palmelloid, dendroid, filamentous, branched and unbranched, heterotrichous, siphonaceous and parenchymatous.
- ☆ Flagella are 1-many, equal in size and inserted either apically or sub-apically.
- ☆ Cell wall made up of cellulose, which comprised of hydroxyproline glycosides or xylans and mannans.
- ☆ Semi-permeable membrane surrounding protoplast. Cytoplasm with small vacuoles.
- ☆ Flagella cells have eye-spot or stigma in the anterior position, which remain inserted in one side of chloroplast.
- ☆ Pigments are located in the chloroplast which has pyrenoids.
- ☆ Main pigment is chlorophyll a and b; with α and β carotenes and xanthophylls.
- ☆ Reserve food material is starch composed of amylose and amylopectin.
- ☆ Reproduce by vegetative (cell division and fragmentation) a sexual (zoospore, alpanospore, akinete *etc.*) and sexual (isogamy to oogamy).
- ☆ Zygospore or oospore only diploid structure in the life cycle.

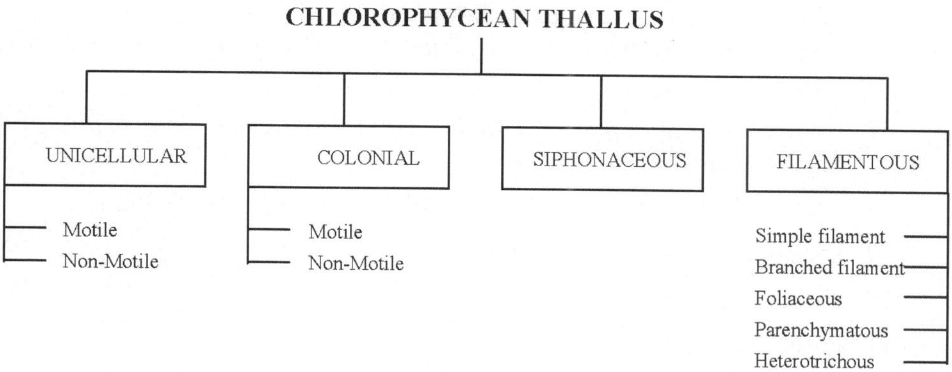

Forms of Algae Thallus Organization in Algae (Plant body types)

Organization of the thallus in algae is basically of two types: unicellular or multicellular. The enormous range of forms exhibited by algae arises from a modification or elaboration of these two types. During its growth cycle, an algae passes through two phases: vegetative (or developmental) and reproductive (or multiplicative). In unicellular algae, the growth in size constitutes the vegetative phase, whereas increase in cell numbers by cell division constitutes the reproductive phase. In multicellular thalli, both the vegetative and reproductive phases undergo cell division and cell growth.

Algae exhibit a great diversity in the organization of the plant body from the simplest motile (*e.g., Chlamydomonas*) or non-motile (*e.g., Chlorella*) unicells to complex forms having several kinds of specialized cells and organs (*e.g., Laminaria,* and *Sargassum*).

Algae sizes vary greatly from microscopic unicellular (*Chlamydomonas, Dufialiella etc.*) attaining a maximum size of 0.5 µm diameter and *Micromonas pusilla* (1-1.5 µm) and species of Chlorella (5-8 µm) to large macroscopic seaweed *Macrocystis pyrifera* reported from South America attaining a length of 213 m (Prescott, 1969), the longest plant in the world. All other members fall in between these limits.

Thallus organization in algae shows a wide range of variation as following habits:

1. Unicellular, *e.g., Chlamydomonas*
2. Colonial (coenobial), *e.g., Volvox, Eudorina,* and *Pandorina.*
3. Palmelloid, *e.g., Chlamydomonas, Tetraspora, Chlorococcus,* and *Palmella.*
4. Dendroid, *e.g., Ecballocystis* and *Prasinocladus.*
5. Coccoid, *e.g., Chlorella, Hydrodictyon, Pediastrum.*

6. Filamentous, *e.g., Spirogyra, Zygnema, Ulothrix, Oedogonium, Nostoc, Oscillatoria, Lyngbya, Anabaena, Tribonema, Cladophora, Pithophora,* and *Bulbochaete.*
7. Heterotrichous, *e.g., Fritschiella, Draparnaldiopsis, Ectocarpus,* and *Stigeoclonium*
8. Siphonaceous, *e.g., Vaucheria, Botrydium, Bryopsis,* and *Codium.*
9. Uniaxial, *e.g., Batrachospermum.*
10. Multiaxial, *e.g., Polysiphonia.*
11. Parenchymatous, *e.g., Ulva, Enteromorpha* and *Sargassum.*

Organisation of Thallus

Salient Features of Chlamydomonas

1. The plant body *of Chlamydomonas* is a unicellular, motile, biflagellate usually egg-shaped thallus.
2. It consists of a tiny protoplast surrounded by a thin delicate but firm cell wall.
3. The cell protoplast is differentiated into a plasma membrane, cytoplasm, neuromotor apparatus, a single nucleus, a green, cup-shaped chloroplast with usually a single pyrenoid, a red eye spot, two contractile vacuoles, ribosomes, perinuclear dictyosomes, mitochondria and endoplasmic reticulum.
4. The *Chlamydomonas* cell is haploid. It reproduces, repeatedly during the growing season by the asexual method of zoospore formation. There is thus a succession of haploid phases.
5. In addition there is production of aplanospores or hypnospores, each of which on liberation behaves like a *Chlamydomonas* cell.
6. Sexual reproduction takes place only once during the season and usually towards the close.
7. The different species of *Chlamydomonas* can be arranged in a series illustrating all the transitional stages in the evolution of sex in green algae from isogamy to oogamy through anisogamy.
8. The diploid fusion cell formed by the union of the gametes in isogamy and anisogamy is, at first, usually a motile quadriflagellate zygote.
9. Sooner or later it loses its flagella, secretes a heavy wall around it to become a zygospore which enters upon a period of rest.
10. The zygospores function in perennation, dispersal and production of haploid swarmers (meiospores).
11. With the return of suitable conditions, the zygospore germinates.

12. During germination the diploid nucleus undergoes meiosis followed by cleavage of the protoplast to form the haploid swarmers.
13. The four resultant haploid daughter protoplasts or swarmers escape by the cracking of the zygospore wall. Each liberated swarmer (haploid) behaves like a *Chlamydomonas* cell.
14. The thick-walled hypnospores, zygospores and to some extent palmella stage serve as means of perennation.
    - ✰ Members of the green algae are well known for the extraordinary variety of forms, shapes and organization of plant body.
    - ✰ The plant body may be single celled or many celled.
    - ✰ In size it varies from minute unicells no more than a micron or two in diameter, to a few feet long, strand like structure.

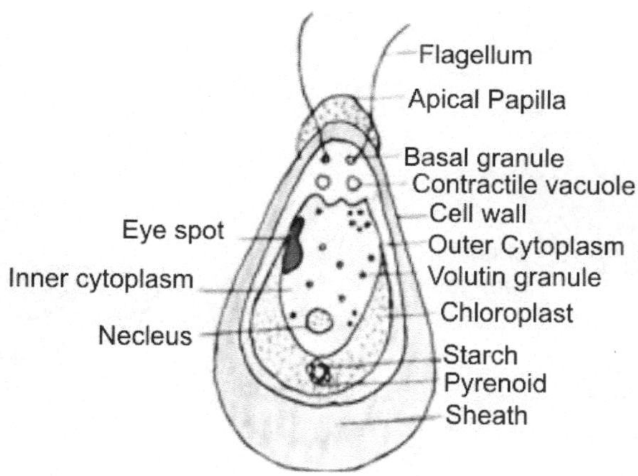

Figure 1.1: *Chlamydomonas* Showing Motile Unicellular Thallus.

- ✰ Morphologically may consist of motile or non-motile cells which are unicellular, motile or non-motile colonies. Simple or branched filaments, siphonaceous or coenocytic filaments, foliaceous thalli, cushion-like thalli and highly evolved heterotrichous filament.
- ✰ However, it is always a simple construction.
- ✰ Shows no differentiation into true root, stem and leaves.
- ✰ For this reason, the plant body of the algae is called a thallus.

Unicellular Thallus

A. Motile Unicellular Thallus

- ✰ The simplest structure is seen under this category where unicellular, spherical, oval or pear-shaped motile plant body may be isolated, individual or colonial form being not a mere assemblage of flagellated cells, but in a coordinate whole behaving as a well-organized unit with a definite number of cells arranged in a specific manner known as a coenobium. Example: *Chlamydomonas.*

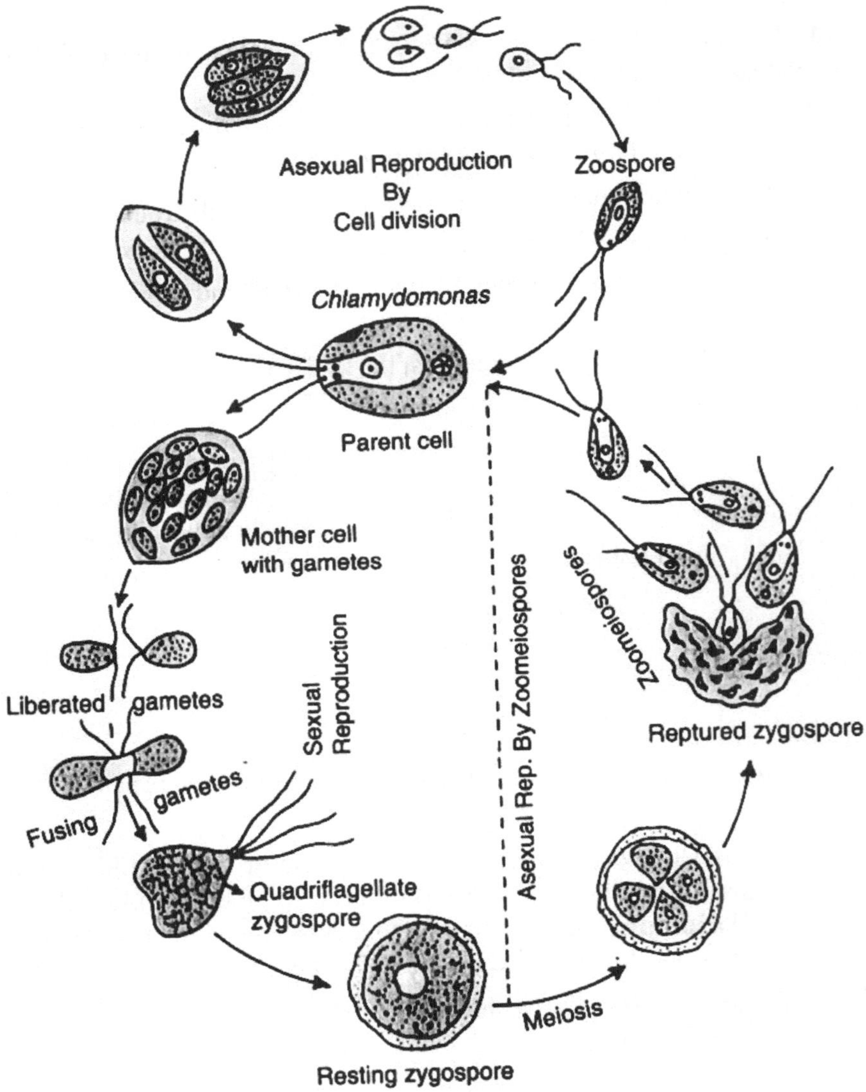

Figure 1.2: *Chlamydomonas* sp. Asexual and Sexual Life Cycle.

- ☆ It is furnished with two flagella at its anterior end and has a single cup-shaped chloroplast situated in its posterior part.
- ☆ The single nucleus lies in the cavity of the cup surrounded by cytoplasm.

- ☆ There are two pulsating contractile vacuoles at its anteiar end and a single red eye-spot.

B. Non-Motile Unicellular Thallus

- ☆ *Chlorella* and *Chlorococcum* represent such type of a thallus.
- ☆ The plant body is a small, spherical cell.
- ☆ It is non-motile and lacks flagella, eye-spot and contractile vacuoles.
- ☆ It has a parietal chloroplast and a centrally located nucleus.
- ☆ Such non-motile unicellular forms are called coccoid chlorophyta.

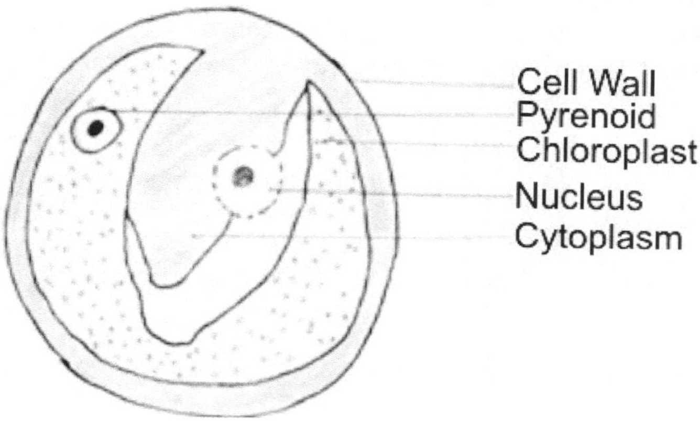

Figure 1.3: *Chlorella* Showing Unicellular Non-motile Thallus.

The family Chlorellaceae includes forms which are usually solitary, free living or symbiotic and few colonial members (*Radiococcus*). The cells are usually spherical or ellipsoid. Reprouction is by autospores. The autospores are non-motile sores resembling the parent in form and structure. The important unicellular member of this family is *Chlorella.*

Chlorella is known for its extreme simplicity. The plant is a unicell which at the most may grow to 10 μm diameter but usually it is much smaller. Most of the cells of *C. variegate* is approximately to 7.5 μm in diameter. The small cells are non-motile, round or oval, usually found solitary, some times in groups. The cell protoplast is enclosed in a membrane which is selective in what it will allow to enter the cell. External to the cell membrane is the thick cell wall. There is a single thin, usually cup-shaped or bell-shaped chloroplast whch is parietal in postion. Sometimes the chloroplast is a curved band or of irregular shape usually flattened agains the cell wall. In the cavity of the chloroplast is the colourless cytoplasm in which lies the single control nucleus. The pyrenoids are usually absent. The stigma and contractile vacuoles are lacking.

Structure Reproduction

The sole method of reproduction is asexual and azoosporic. The contents of the cell divide into 2, 4 (B), 8(C) sometimes 16 daughter protoplasts. Each daughter protoplast rounds off to form a non-motile spore. These autospores (spores having the same distinctive shape as the parent cell) are liberated by the rupture of the parent cell wall (D). On release each autospore grows to become a new individual. The presence of sulphur in the culture medium is considered essential for cell division. It takes place even in the dark with sulphur alone as the source material but under light conditions nitrogen is also required in addition. Pearsal and Loose (1937) reported the occurrence of motile cells in *Chlorella.* Bendix (1964) also observed that *Chlorella* produces motile cells which might be gametes. These observations have an important bearing on the concept of the life cycle *of Chlorella* which, at present, is considered to be strictly asexual in character.

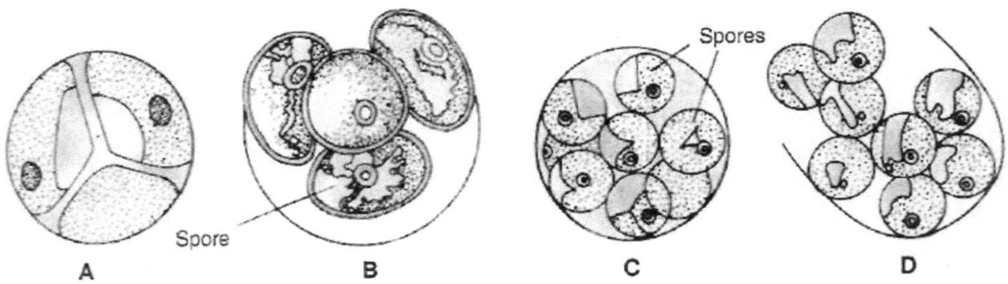

Figure 1.4(A – D): *Chlorella.* Asexual Reproduction.

Asexual reproduction in *Chlorella ellipsoides* has been studied in detail and the following four phases have been observed during the asexual reproduction (Figure 1.4).

(i) *Growth phase.* During this phase the cells grow in size by utilising the photosynthetic products.

(*ii*) *Ripening phase.* In this phase the cells mature and prepare themselves for division.

(*iii*) *Post ripening phase.* During this phase, each mature cell divides twice either in dark or in light. The cells formed in dark are known as *dark nascent cells* which in turn give rise to photosynthetically active cells. During transition from dark to light phase, cells again grow in size.

(iv) *Division phase.* During this phase the parent cell wall ruptures and unicells are released.

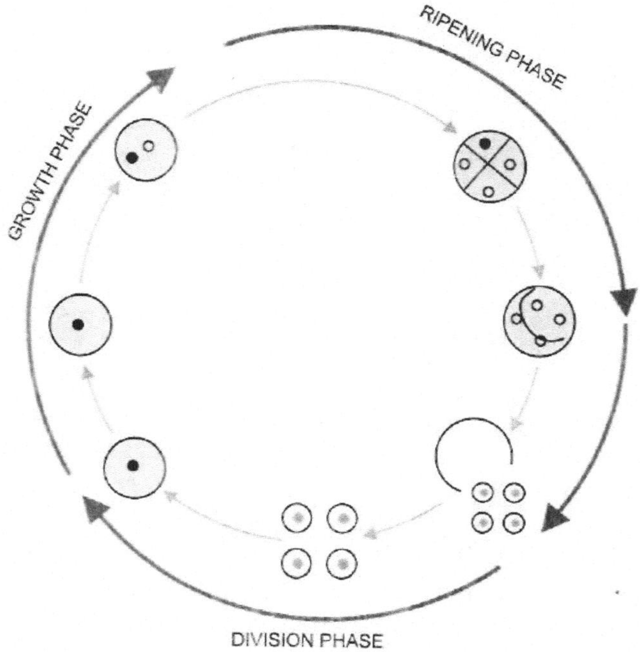

Figure 1.5: Phases in the Life Cycle of *C. ellipsoides.*

Colonial Thallus

Many of the green algae have a thallus consisting of a loose assemblage of cells mechanically held together generally in a gelatinous envelope. It is known as a colony. The cells in the colony have little or no dependence on each other.

A. Motile Colonial Thallus

- ☆ Consists of a definite number of motile, unicellular *Sphaerella* types of cells held together in a common mucilageneous sheath.
- ☆ Individual cells complete and independent of each other.
- ☆ Act together and bring about movement of the colony.

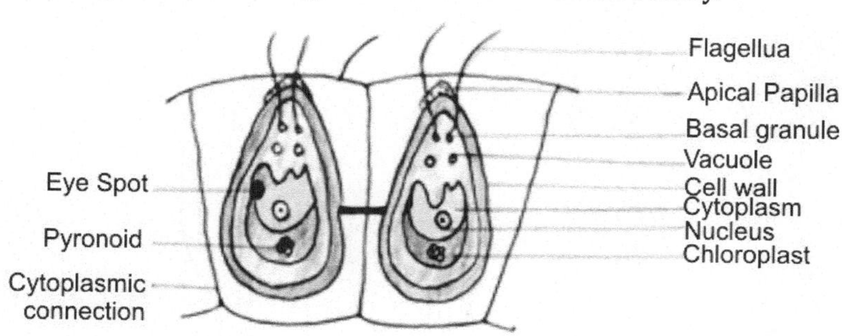

Figure 1.6: Cell Structure of *Volvox* Showing Motile Colonial Thallus.

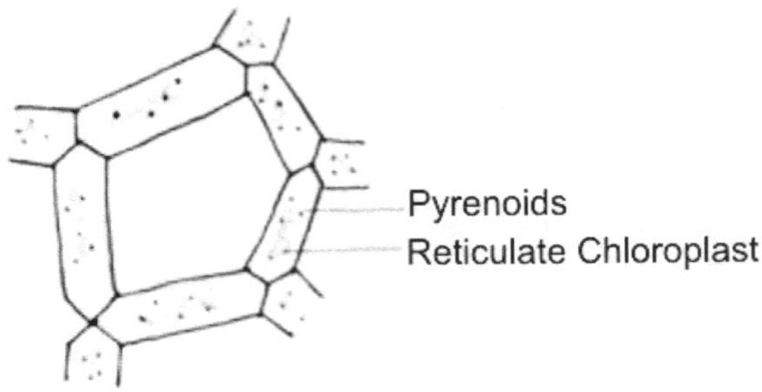

Figure 1.7: Colony of *Hydrodictyon* Showing Non-motile Colonial Thallus.

- Example: Colony of *Volvox* represents most highly developed motile colonial thallus. Made up of thousands of cells connected together by cytoplasmic strands.

B. Non-Motile Colonial Thallus

- In certain members of the chlorococcales, the small non-motile cells are held together to form non-motile colonies with either a definite or an indefinite number of cells. They are free-floating colonies.
- Example: *Hydrodictyon* represents the highest state of organization of this type.

Siphonaceous Thallus

- Also called coenocytic thallus.
- Uni-cellular thallus is enlarged to form a non-septate multi-nucleate sac-like or tubular-structure which is not divided into cells, in the somatic phase.

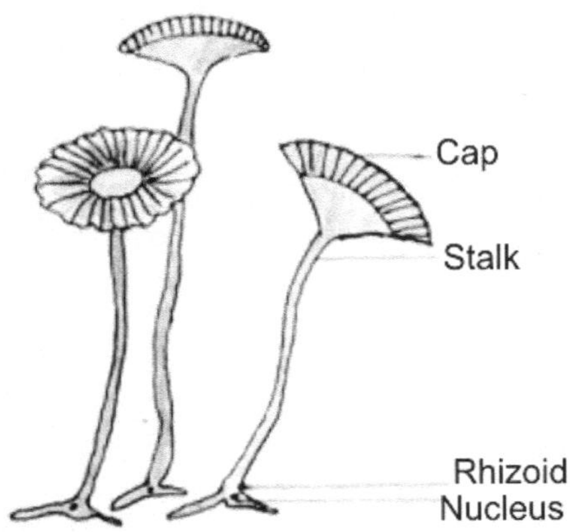

Figure 1.8: *Acetabularia* Showing Early Stage of Siphonaceous Thalli.

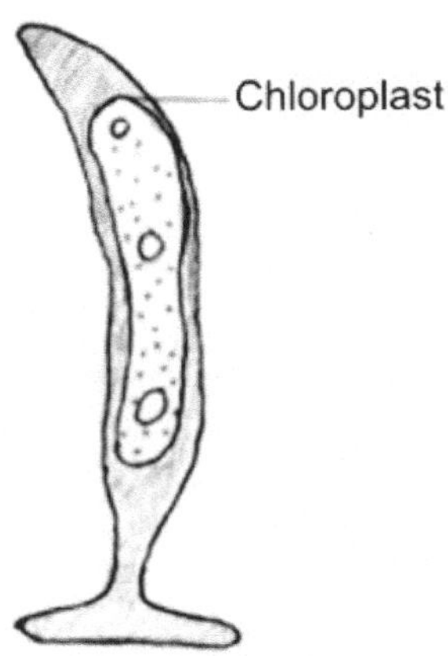

Figure 1.9: *Charcium* Showing Tabular Uninucleate Thallus.

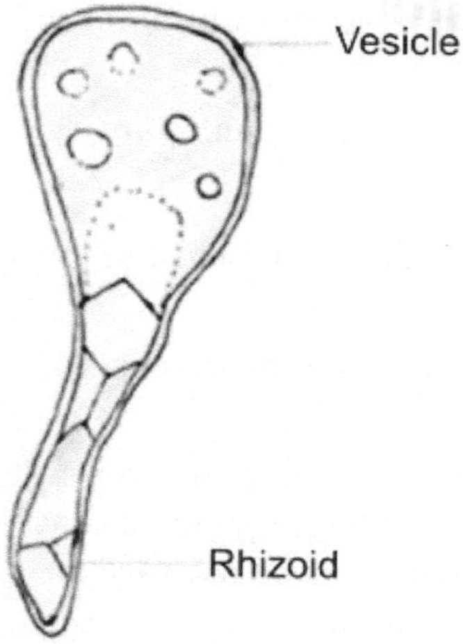

Figure 1.10: *Protosiphon* Showing Unseptate Branched, Tabular, Thallus.

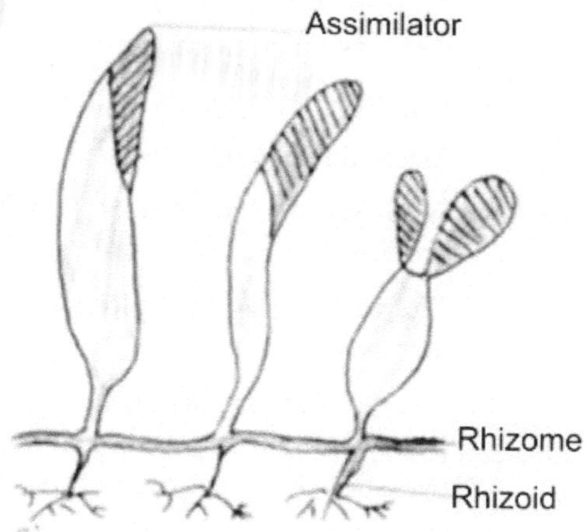

Figure 1.11: *Caulerpa* Marine Sphonaceous Form.

- ☆ *Acetabularia* and *Characium* represent the early stage in the development of a Siphonaceous body. They are both uni-cellular green algae. Former is tubular and the latter is umbrella shaped.
- ☆ *Acetabularia* consists of a talk ending in a umbrella-like cap. It consists of a single nucleus which is present at the base of the stalk.
- ☆ In both cases, as the unicell reaches maturity and enters the reproductive phase, nucleus undergoes division to form a number of nuclei.
- ☆ The tendency to become multinucleate for a short time becomes more permanent in *Protosiphon* which has unseptate branched, tubular thallus containing numerous nuclei.
- ☆ In submerged Siphonaceous form, buoyed up by water, the coenocytic thallus is more elaborate. Example: As in *Caulerpa*. It is a tropical marine genus in which the thallus extends to a foot or so in length. It is differentiated into a creeping structure resembling a rhizome.
- ☆ The rhizome gives rise to root like holdfast from its underface and erect leafy shoots from its upper face.

Salient Features of *Caulerpha*

1. The coenocytic thallus simulates to a remarkable degree, the external appearance and size of a vascular plant with a creeping root stock.
2. Owing to the absence of septa the multinucleate lining layer of cytoplasm containing numerous, minute, round nuclei and disciform chloroplasts and the single, large central vacuole is continuous throughout the plant.

3. Numerous cylindrical skeletal strands traverse the central vacuole. They serve as mechanical supports.
4. Vegetative propagation is by means of fragmentation.
5. Asexual reproduction by spores is unknown.
6. Sexual reproduction is generally anisogamous and most of the species are dioecious.
7. The thallus being diploid, meiosis takes place at the time of gamete formation.
8. The gametes are motile uninucleate and bifiagellate structures. They are pyriform in shape and each has a single chloroplast and a conspicuous eye-spot. They are distinguishable into micro- and macrogametes.
9. They escape through the ends of extrusion papilla in a mass of mucilage.
10. The fertile portion of the thallus dies soon after the liberation of gametes.
11. The zygote retains all the four flagella for a time. Soon it resorbs the fiagella, rounds off, and secretes a wall around it.
12. Because of gametangial meiosis the life cycle of *Caulerpa* is described as haplobiontic diploid or diplontic.

Multicellular Filamentous

- ☆ In some species of green algae cells are arranged in linear rows called the threads or filaments.
- ☆ Filament is the result of repeated cell division of a non-motile cell in a single plane.
- ☆ The divisions are all in parallel plane resulting in the formation of a chain of cells.
- ☆ The daughter cells in the chain remain united and further keep on dividing in the same plane forming a simple filament.
- ☆ It is the forerunner of various branched types.
- ☆ The filamentous thallus is well adapted for a great variety of fresh water and marine habitats.
- ☆ They are of the following types:
  a. Simple filament
  b. Branched filament
  c. Foliaceous thallus
  d. Massive parenchymatous thallus
  e. Heterotrichous filament

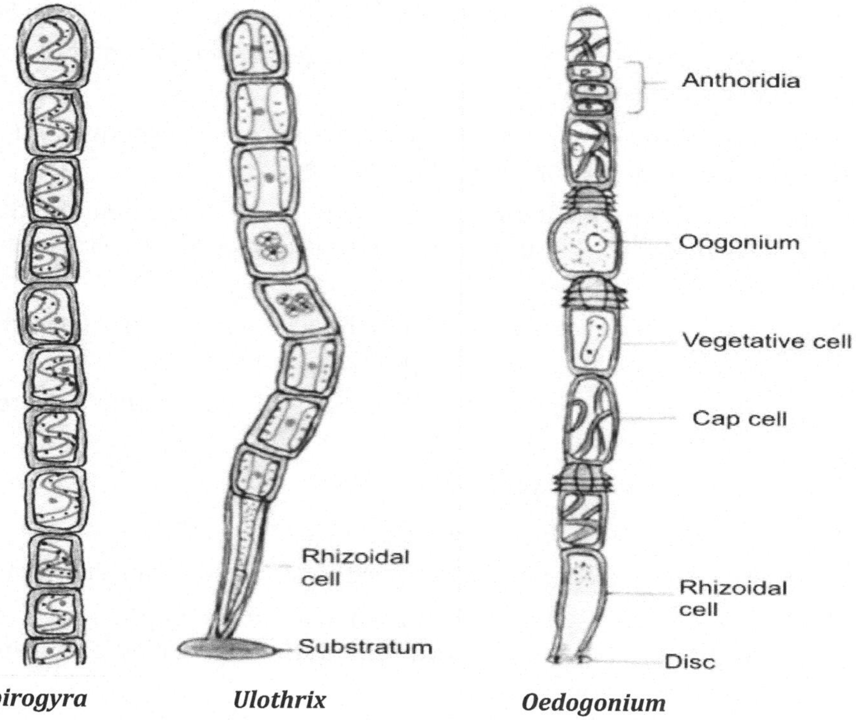

Figure 1.12

Salient Features of *Ulothrix*

1. The thallus is an unbranched filament consisting of short, cylindric or quadrate uninucleate cells.
2. The filament shows distinction into base and apex.
3. The basal, colourless rhizoidal cell is longer and narrower than the others in the filament. It functions as an organ of attachment. All others are green, cylindric or quadrate and similar in all respects. The cell at the free end, however, is dome-shaped rather than cylindrical.
4. The filament illustrates the beginning of specialisation leading to interdependence of cells.
5. The green vegetative cell is made up of the cell wall surrounding the protoplast.
6. The cell protoplast is differentiated into cytoplast, a nucleus, a single large chloroplast shaped like a bracelet and a central vacuole.
7. The chloroplast is girdle-shaped and lies in the middle of the cell surrounding the bulk of the cytoplasm and nucleus. It is thus parietal in position and contains one or more pyrenoids.

8. The haploid filament reproduces both sexually as well as asexually.
9. Asexual reproduction takes by the vegetative method of fragmentation and sporulation.
10. The spores are chiefly zoospores sometimes aplanospores. Both serve to reduplicate the haploid *Ulothrix* filament.
11. Sexual reproduction is accomplished by the formation of biflagellate gametes which like the spores may be produced in any cell of the filament except the basal cell.
12. The fusing gametes are shed into the surrounding water through a pore in the lateral wall so that fertilisation is external.
13. The fusing gametes exhibit functional disparity. One gametes from different filaments fuse. *Ulothrix* thus is dioecious.
14. The thick-walled zygospore is a resting spore.
15. On germination the zygospore produces 4-16 meiospores.
16. On liberation each meiospore produces a haploid *Ulothrix* filament.
17. The zygospore with the contained meiospores may be regarded as the forerunner of the sporophyte generation and called the incipient or primitive sporophyte.
18. Zygospore thus has a three-fold function : *perennation, dispersal* and *production of meiospores.*
19. The simple life history of *Ulothrix* consisting of a haploid plant body (filament) with the zygote as the diploid stage is called haplontic. It is characterised by zygotic meiosis.

Salient Features of *Spirogyra*

1. It is free living green alga that usually occurs floating in standing water in the form of bright, green silky masses slimy to touch.
2. The thallus is filamentous. The filament is a long hair-like unbranched thread. It shows no distinction into base and apex.
3. The filament consists of long cylindrical cells placed end to end in row.
4. All the cells in the filament are similar in structure, function and are independent of each other. There is thus neither any differentiation of cells in the filament nor any division of labour.
5. Each cell consists of a cell wall enclosing the protoplast.
6. The cell wall consists of two concentric layers. The inner is cellulose in nature. The outer is a pectose layer covered with a mucilage sheath. The latter gives a slimy feel.

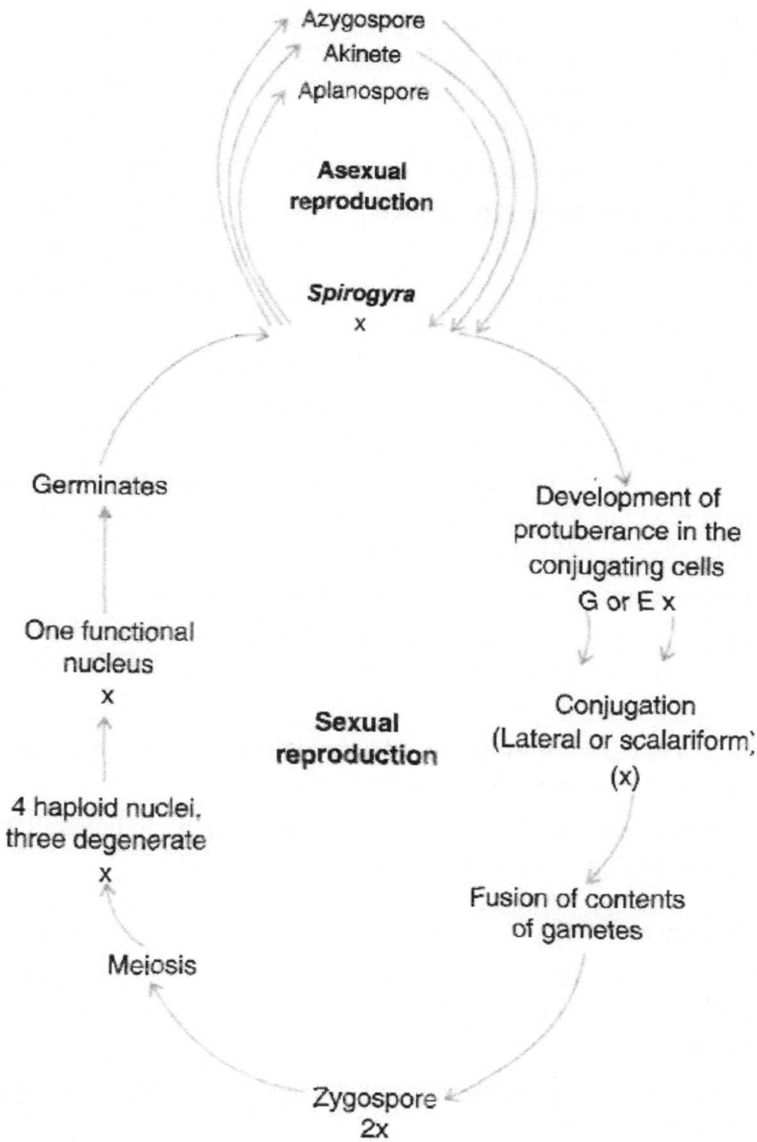

Figure 1.13: Word Diagram of the Life Cycle of *Spirogyra*.

7. The protoplast is differentiated into a plasma membrane, cytoplasm, nucleus, one or more chloroplasts with pyrenoids and a central vacuole.
8. The cytoplasm forms a utricle around the large, central vacuole containing the cells sap.
9. The single large nucleus surrounded by a tiny mass of cytoplasm remains suspended in the centre of the vacuole by a number of cytoplasmic strands.

10. The number of chloroplasts varies from one to several. Each chloroplast is a green spiral band and is parietal in position. It is embedded in the cytoplasm and coils round from one end of the cell to other in an anticlockwise direction.
11. Each chloroplast bears a number of pyrenoids equidistant of each other.
12. The filament during the growing season multiplies by the method of fragmentation.
13. Asexual reproduction is isogamous and accomplished by the method of conjugation which may be scalariform or lateral.
14. In the former, conjugation tubes connect the young and recently divided opposite cells of the two conjugating filaments.
15. The gametes are non-motile, similar in size and structure but differ in their behaviour. The male gamete migrates through the conjugation tube by amoeboid movements and joins the passive female gamete in the opposite cell to form a zygospore.
16. Lateral conjugation is of two types, indirect and direct.
17. In indirect lateral conjugation male gamete migrates through the conjugation canal formed between the adjoining cells of the same filament.
18. In direct lateral conj ugation the male gamete migrates through a pore in the intervening cell wall between the two adjacent cells.
19. The zygospore is a drought resisting structure. It germinates after a period of rest and develops directly into a new filament.

Simple Filament

- ☆ Simple multicellular forms like *Spirogyra* have a plant body in which the cells are arranged from end to end in a single file and are held together firmly.
- ☆ It is a long., thread like, unbranched, elementary type of multicellular thallus.
- ☆ Like colonial thallus, all the cells are alike, self sufficient and independent of each other.
- ☆ Because of the simple filament of *Spirogyra*, it is regarded as an advanced type of colonial thallus in which the cells are arranged in a row. It is called filamentous colony.

Ulothrix has an unbranched filament. The simple filament of it consists of similar cells but it is attached to the substratum at one end by a rhizoidal cell specically modified for this purpose.

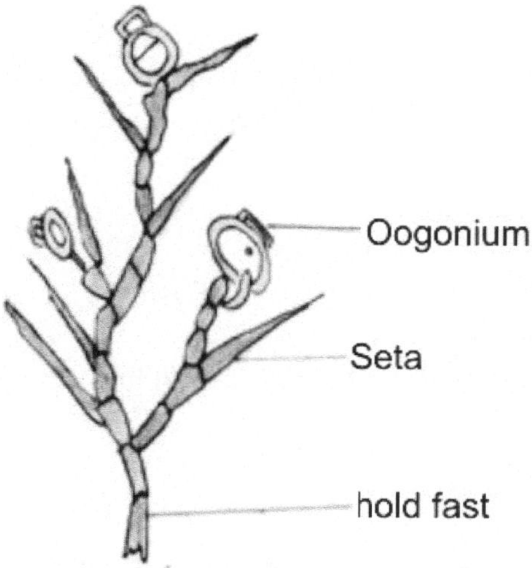

Figure 1.14: ***Bulbochaete*** **Showing Branched Filamentous Thallus.**

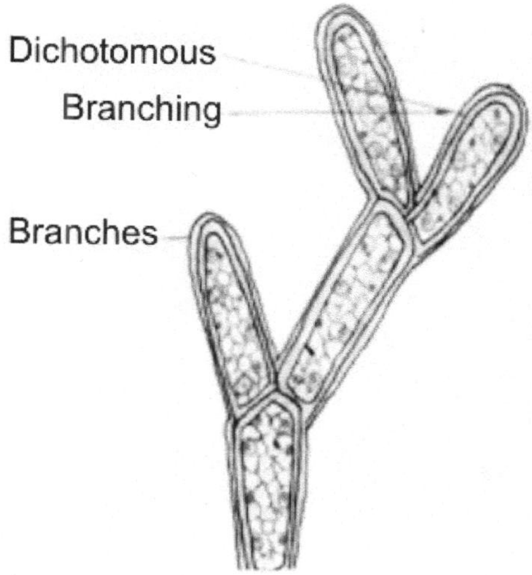

Figure 1.15: ***Cladophora*** **Showing Branched Filamentous Bushy Thallus.**

Oedogonium is an example of simple multi-cellular filament with a branched or lobed rhizoidal cell. Besides the rhizoidal cell, there are green vegetative cells for nutrition cap cells for cell division, zoospore formation and highly specialized reproductive cell which function as sex organs.

Figure 1.16: ***Ulva fasciata*** **Showing Foliaceous Thallus.**

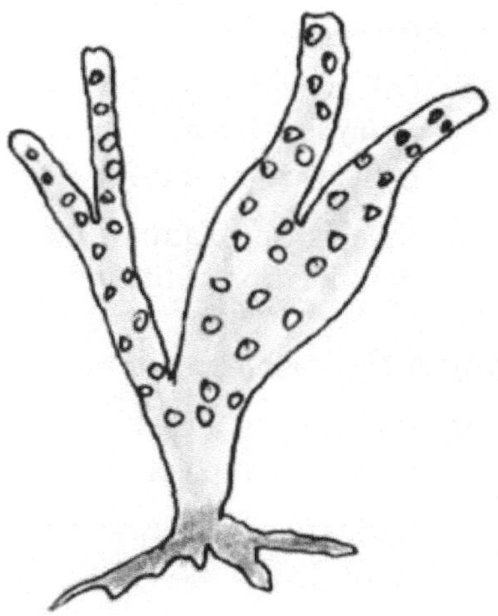

Figure 1.17: ***Ulva reticulata*** **Showing Foliaceous Thallus.**

Branched Filament

- ☆ More advanced green algae such as *Bulbochaete* and *Chaetophara*, the thallus is a branched filament.
- ☆ Formed when occasional cell in the filament divides in a second plane.

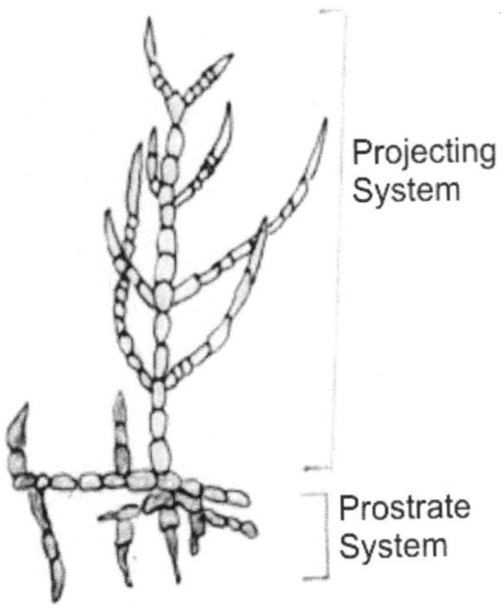

Figure 1.18: ***Stigeoclonium*** **Showing Heterotrichous Thallus.**

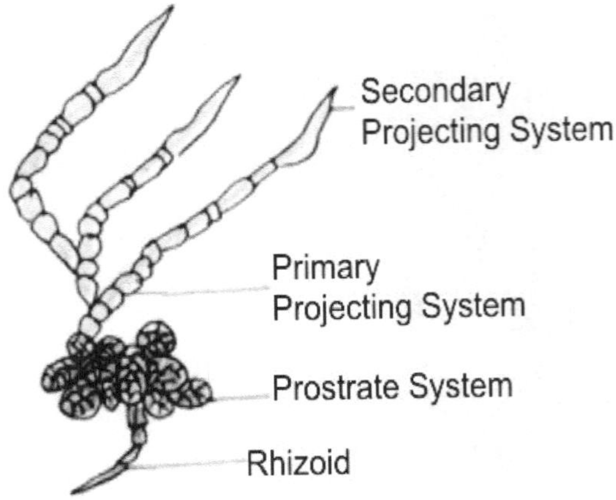

Figure 1.19: Heterotrichous Thallus of ***Fritschiella*** **with Greater Elaboration of Projecting System.**

- Branches arise as lateral outgrowths from the same filament.
- The branched filament of *Cladophora* forms a more tufted and bushy thallus.

Foliaceous Thallus

- Cell divisions occur in two plane (transverse, longitudinal) in the cells of a simple filament, the result in a thin, flat, plate like thallus resembling papery expansions.
- *Ulva* is an example having a foliaceous type of thallus which is a modification of filamentous habit.

Massive Parenchymatous Thallus

- It occurs as a result of divisions in a plane parallel to the surface in a foliaceous type of thallus.
- In this way, a massive bulk thallus, two to three or more cells in thickness is formed.

Heterotrichous Filament

- Most advanced and evolved type of thallus met within the chaetophorales among the green algae.
- Basal portion of the thallus creeps along the substratum and is called the prostrate system.

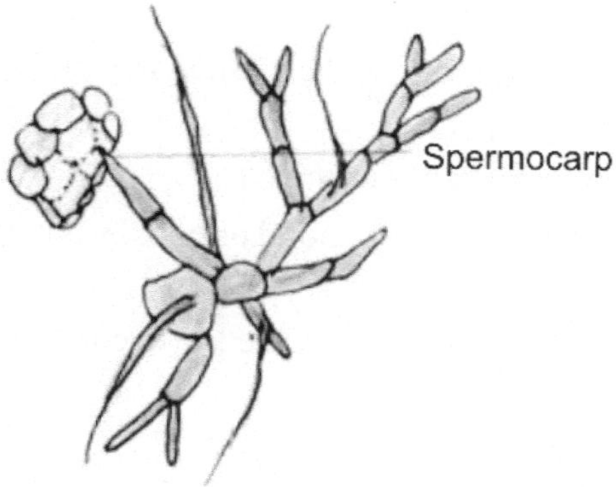

Figure 1.20: ***Coleochaete.***

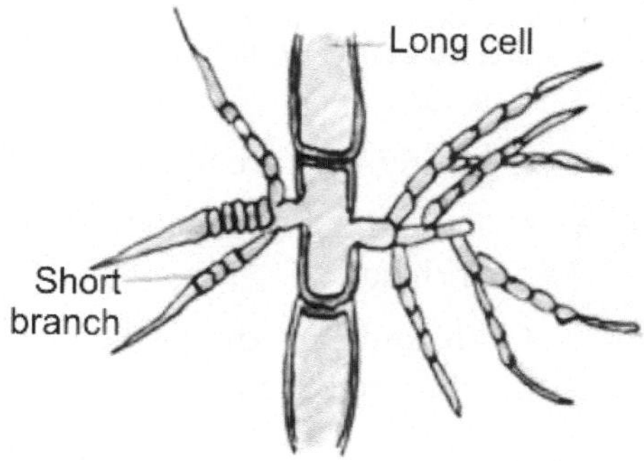

Figure 1.21: ***Draparnaldia.***

- ☆ From it, arise the upright filament constituting the erect system.
- ☆ In *Stigeoclonium*, both the prostrate and erect systems are equally well developed.
- ☆ Prostrae system of *Fritschiella* consists of clusters of rounded cells, attached to substratum by rhizoids.
- ☆ From this, arise a number of upright branched threads consisting of very short cells. These constitute the primary projecting system. Above the later is the secondary projecting system comprising of elongated cells with branches.
- ☆ The disappearance of erect system has resulted in the discoid forms of species of *Coleochaete.*
- ☆ *Draparnaldia* is an example of disappearance of prostrate system and elaborate development of the erect system. Prostrate system is vestigial and replaced by hold fast. Erect system bears lateral branches from which arises whorls of branched laterals, at short intervals. There is thus differentiation into nodes and internodes.

Charophyceae

Chara: Valliant, 1919

Systematic Position

Class	: Charophyceae
Order	: Charales
Family	: Characeae
Genus	: *Chara*

Characteristic Features

1. The genus *Chara* has about 90 species, is widely distributed in standing waters.
2. Mostly, they are fresh water forms, a few are marine forms.
3. The plant is attached to the muddy bottom of the pond or pool by means of rhizoids.
4. Rhizoids are thread like, uniseriate, multicellular, well-branched and white coloured strcutures.
5. These arise from the lower nodes of the main axis.
6. Oblique septa are present in rhizoids.
7. The plant is distinguished into internodes and nodes.
8. The internode is constituted by long cylindrical, cortical cells which are much smaller in diameter.
9. At the nodes of the leaves, single-celled branches occur and they are known as secondary laterals or stipules.
10. Single nucleus and several chloroplasts are present in the dense cytoplasm, chloroplast lacks pyrenoids.

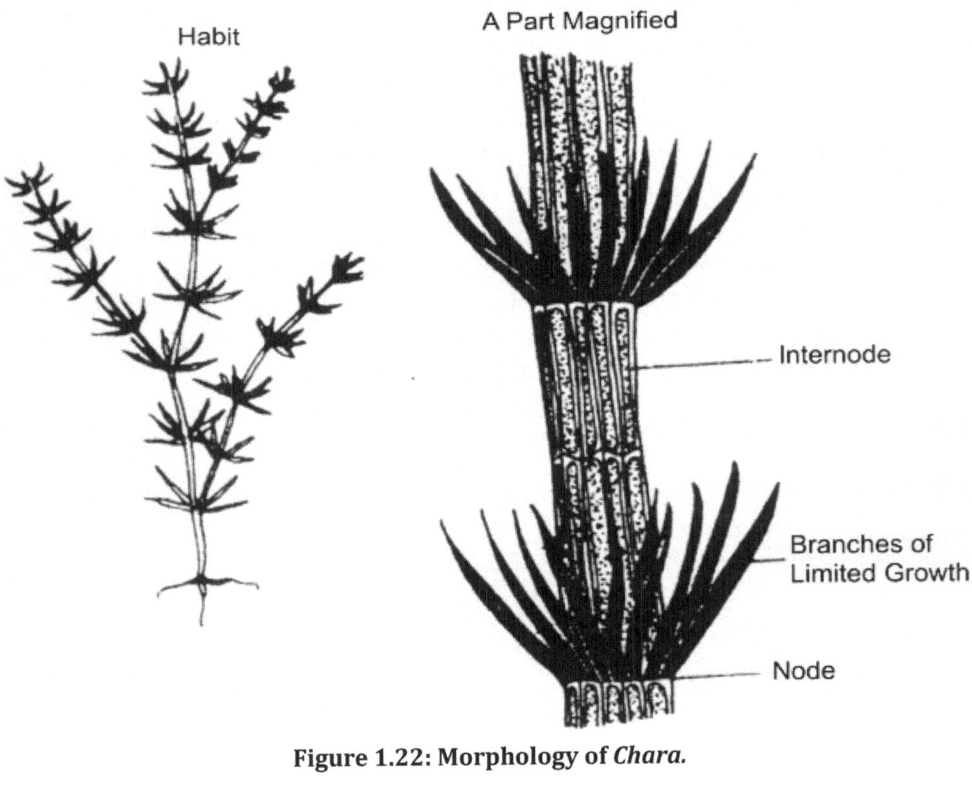

Figure 1.22: Morphology of *Chara*.

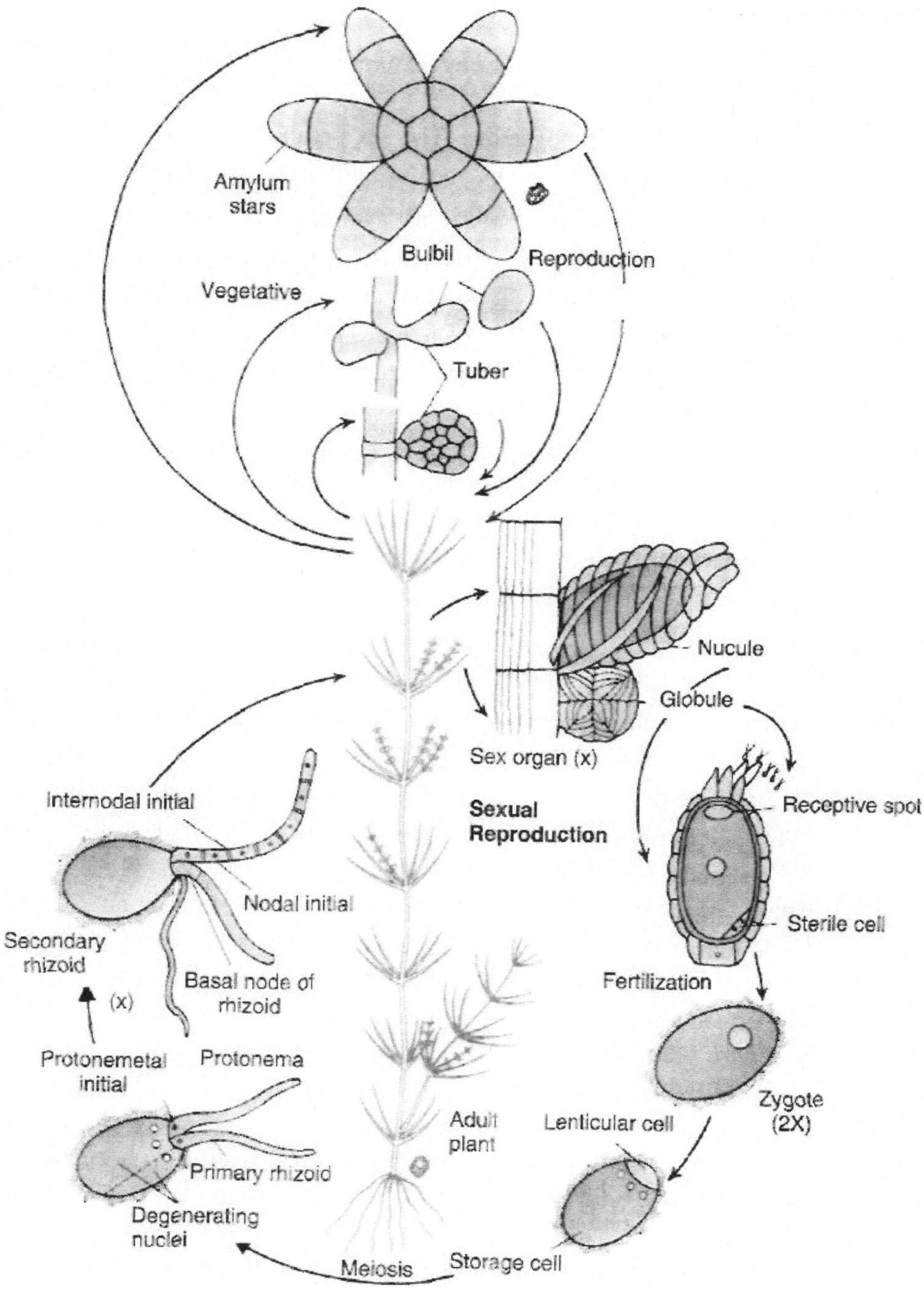

Figure 1.23: Diagrammatic Life Cycle of *Chara*.

11. *Chara* reproduces by vegetative and sexual methods asexual reproduction is entirely absent.

Points to Remember

1. The thallus shows an elaborate organisation and may attain a height of 20 to 30 cms. It has the appearance of miniature horse-tails (*Equisetum*).
2. The main axis consists of a series of alternating nodes and internodes.
3. The interaode consists of a single, undivided, elongated, cylindrical cell.
4. The node is made up of a transverse layer of short cells.
5. From the nodes of the central axis arise whorls of primary laterals of limited growth. The primary laterals are also divided into a few nodes and internodes. From their nodes may arise the secondary laterals.
6. From the axil of the primary laterals arise one or two long branches which continue the growth of the thallus.
7. The internodal cells in the corticated species are surrounded by a number of sheath cells arising from the nodes above and below.
8. The growth in length both of the main axis and the branches takes place by means of a single, dome-shaped apical cell.
9. Asexual reproduction by spores is entirely absent.
10. Vegetative reproduction takes place by the formation of amylum stars, bulbils and secondary protonemata.
11. Sexual reproduction is oogamous. The sex organs are large, highly specialised and complicated in structure.
12. The orange or red antheridia and the oval oogonia lie in pairs. The oogonium always lies above the antheridium at the fertile nodes of the short, primary laterals in the monoecious species.
13. The oval oogonium is surrounded by a protective sheath. It consists of five long spirally wound cells ending either in a five celled corona (*Chara*)/ Chara or ten-celled (*Niiella*).
14. The oogonium produces a large uninucleate egg.
15. The antheridium consists of a wall composed of eight long, curved, plate-like cells, the shield cells. The shield cells enclose an internal cavity. From the middle of each shield cell arises a rod-like manubrium. It bears at its inner end capitulla cells ending in long, whip-like, branched or unbranched spermatogenous filaments. The latter fill the antheridial cavity with a dense tangle.
16. Each spermatogenous filament consists of 5 to 200 short, discoid sperm mother cells.
17. The egg is retained within the oogonium and fertilization is internal.
18. Each spermatozoid mother cell gives rise to a single biflagellate sperm.

19. The zygote or the oospore germinates after a resting period.
20. Prior to germination the zygote nucleus divides to form four daughter nuclei. It is suggestive of zygotic meiosis. Three of the haploid nuclei gradually degenerate.
21. The *Chora* plant is thus haploid. The only diploid structure in the life cycle is the dormant zygote.
22. The zygote, on germination, produces the haploid protonema. From it the *Chora* plant arises as a lateral bud-a feature approaching some of the bryophytes (mosses).

Nitella: Agardh, 1824 (A Little Star)

Systematic Position

Class : Charophyceae

Order : Charales

Family : Characeae

Sub Family : Nitelloideae

Genus : *Nitella*

Characteristic Features

1. *Nitella* occurs in shallow water along the edges in pools, lakes and slow flowing streams attached to the sandy or muddy substratum.
2. The plant has a jointed, central main axis with a whorl of branches arising from each joint. It is anchored to substratum of multicellular, branched, colourless, rhizoids.
3. The main axis is differentiated into a series of well, marked attending nodes and internodes. The internodes is a single, elongated, individual cylindrical cell several timer longer than broad without enshething layer of Cortex (ecorticate).
4. The more fluid, less denser endoplasm is in a state of constant rotation. This streaming movement of endoplasm is termed cyclosis.

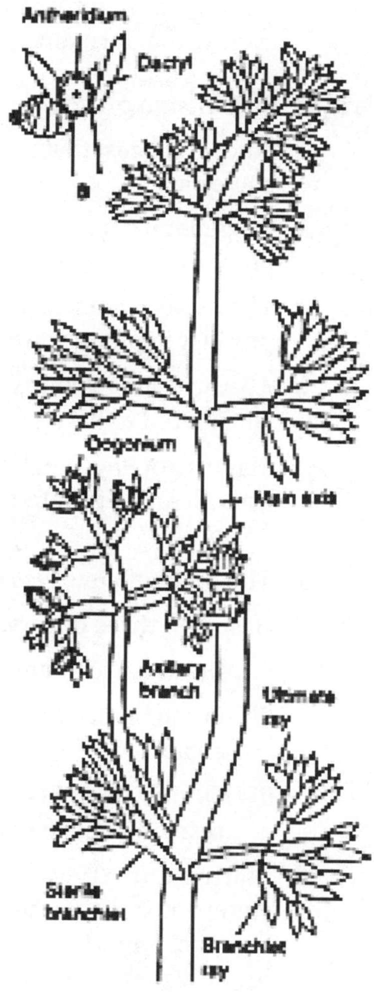

Figure 1.24

5. The main axis and the auxillary branches show indefinite growth by the activity of a dome shaped apical cell located at the tip.
6. Sexual reproduction is oogamous. The sex organs are large, highly specialized and complicated in structure.
7. Antheridium consists of a jacket layer of eight closely fitting shield cells. From the middle of each arises elongated cell called the manubrium.
8. The manubrium at its distal end bears a terminal cell known as primary capitulum which buds about six secondary capitula.
9. The diploid zygote or the oospore germinates after a resting period to produce the haploid protonema.
10. The *Nitella* plant is thus haploid; the only diploid structure in the life cycle is the dormant zygote.

Origin of Chlorophyta

- ☆ Studies suggest that green alga developed from a group of photosynthetic bacteria.
- ☆ Progenetors possessed chlorophyll a and β-carotene as the principal photosynthetic pigments.
- ☆ They used water as a source of hydrogen as released oxygen as a by-product in photosynthesis.
- ☆ All the other groups except Rhodophyceae and Cyanophyceae are considered to have evolved from the Chlorophyceae.
- ☆ Majority of Phycologists hold that green algae have evolved from simple, unicellular, flagellated ancestors possibly similar to some of the present volvocales.
- ☆ The point of view was first put forward by Blackman, West (1900).
- ☆ Based on three observations namely:
  - i) Occurrence of unicellular forms in the chlorophyta.
  - ii) Frequent production of a unicellular stage by the multi-cellular green algae.
  - iii) Similarity of this stage in the different species of green algae and its resemblance to unicellular volvacales.
- ☆ Mattox and Stewart (1977) visualized the green walled flagellate Hetero-matrix as the extant remnants of the early green algae which are considered to have undergone extensive evolutionary pathways and diversification in the cellular organisations to have given rise to Chlamydomonas and to the entire range.
- ☆ The advanced chlorophytes arose from the primitive unicellular flagellate ancestors along three separate evolutionary lines of development.

- ☆ In one of these evolutionary pathways the primitive motile unicellular ancestars retained their motility and became grouped in Coenobia. This is called the Volvocline Line.
- ☆ The order Volvocales represents the Volvocline line.
- ☆ In the second series, the flagellate ancestors lost both motility and capacity to divide into somatic cells resembling the parental cells.
- ☆ It is the Chlorococcine line, illustrated by the Chlorococcales, and the coenocytic orders Caulerpales, Dasycladales and Siphonocladales.
- ☆ The third evolutionary trend known as the Tetrasporine line involves the loss of motility and retention of the capacity of these cells to divide directly into somatic cells resembling the parent cells.
- ☆ This evolutionary trend is present in the present day Tetrasporales.

Evolutionary Tendencies

The chlorophycophyta exhibit both primitive and advanced structural features and methods of reproduction.

- ☆ Origin of Filamentous form from motile unicellular ancestor.
- ☆ Multicellular condition in green algae arose through the adherence of daughter cells after cell division.
- ☆ Loss of motility in ancestral unicell in the vegetative condition with retention of ability to divide freely.
- ☆ Division occur in single plane and daughters remain united.
- ☆ The result was a row of cells called simple filament Ulothrix.
- ☆ Transverse division of cells derived from longitudinal division of flagellates via forms like *Chlamydomonas seriata*.
- ☆ First division of this alga is oblique.

Figure 1.25: Oblique Division.

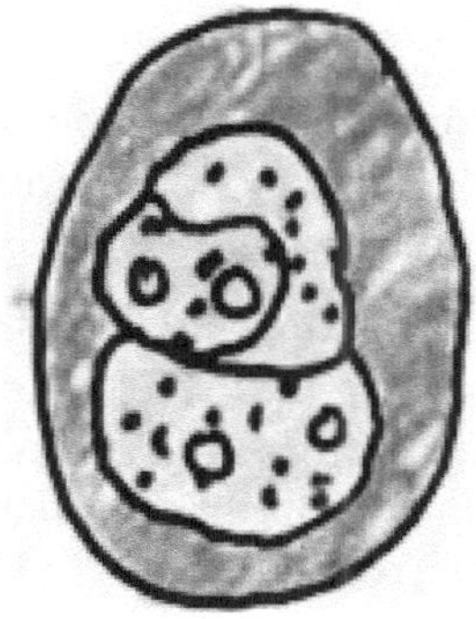

Figure 1.26

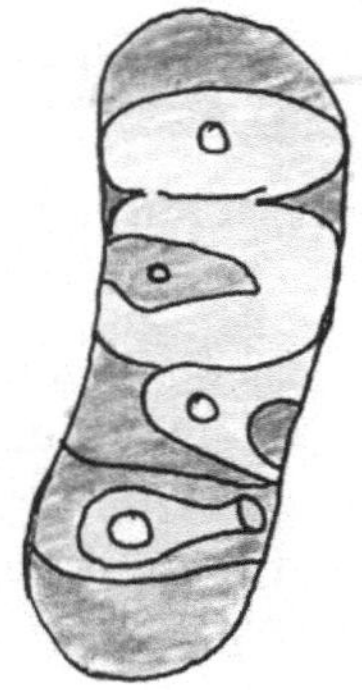

Figure 1.27

- The two daughter protoplast after division, rotate so that the division comes to transverse with the parent cell.
- The two daughter cells again divide longitudinally producing a row of 4 daughter cells. This arrangement is significant in understanding the origin of filamentous habit in algae, from unicellular ancestor.
- This hypothesis was supported by Fritish (1935) who cited as evidence the early growth of many filamentous green algae.

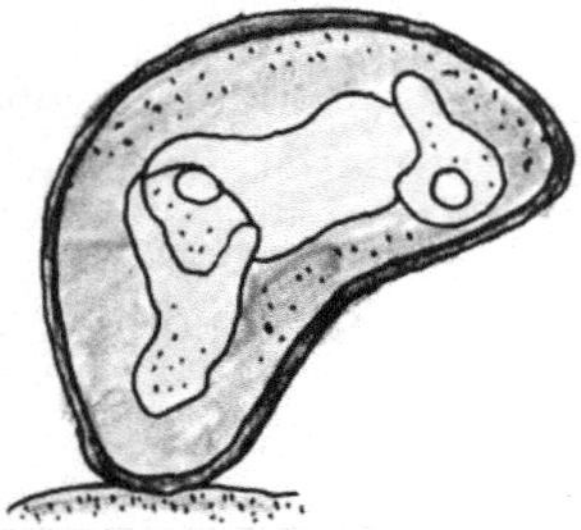

Figure 1.28

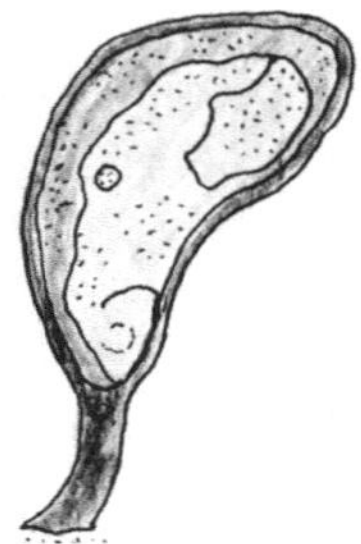

Figure 1.29

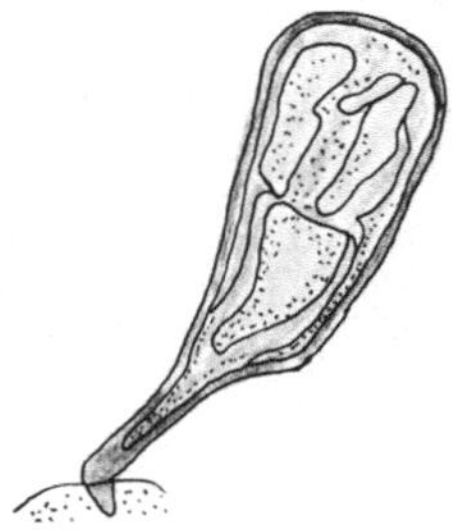

Figure 1.30

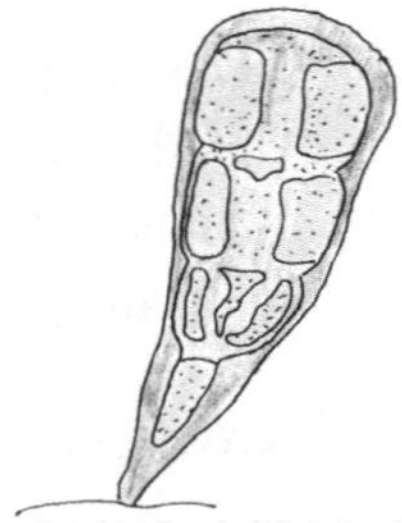

Figure 1.31

- In *Ulothrix, Oedogonium* as many others, the unicellular motile reproductive cells such as the zoospore comes to rest on its base after a short period of time and activity.
- It retracts its flagella and attaches itself to the substratum. In the quiescent condition, it secretes a thin cellulose wall around it.
- Its protoplast turns to 90° so that its axis coincides with the transverse axis of the filament which arises from it.
- The quiescent – zoospore protoplast then divides into two daughter cells which remain united.
- With further divisions in the same plane a filament of cells (multicellular condition) is produced.
- Behaviour of zoospore at the type of germinaton recapitulates the evolution of the filament. Whether the unicellular ancestor was motile *Chlamydomonas* type or non-motile *Chlorella*, the opinion is divided.
- There is a third suggestion that holds the via media. According to this, filamentous form with uninucleate cells like *Ulotrichales, Ulvales* and *Chaetophorales* evolved from the motile unicellular ancestor and the filamentous form with multicelluleate cells such as *Cladophorales, Siphonocldales, Dasycladales* and *Siphonales* originated from the non-motile unicellular chlorococcales such as *Chlorococcum*.

☆ The *Oedogoniales* and *Conjugales* diverged early from the main stream of evolution (tetrasporine line) as represented by the Ulotrichales, Ulvales and Chaetophorales.

Origin of Filamentous (Multicellularity) from Palmelloid

☆ Filament evolved though the aggregation of separate cells into a colony.

☆ Unicellulor motile alga like *Chlamydomonas* loses motility under certain unfavourable conditions.

☆ Several individuals in this condition become aggregated in a mass of mucilage.

☆ This temporary association in *Chlamydomonas* in individuals is called palmella stage.

☆ Blackman (1900) advocated the origin of filamentous form in the ulotrichales from the palmelloid forms. As example he cited examples of *Palmodictyon varians* in which cells are linear.

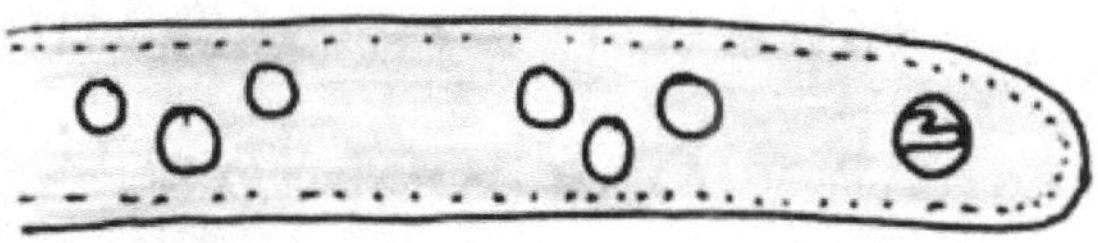

Figure 1.32: *Palmodictyon varians.*

☆ He assumed that the restriction of cell division in such forms to a single transverse plane led to forms like *Radiofilum* and *Geminella interrupta*, members of Ulotrichales.

☆ Cells are arranged in a row to form a simple filament but they are separated from each other by gelatinous materials.

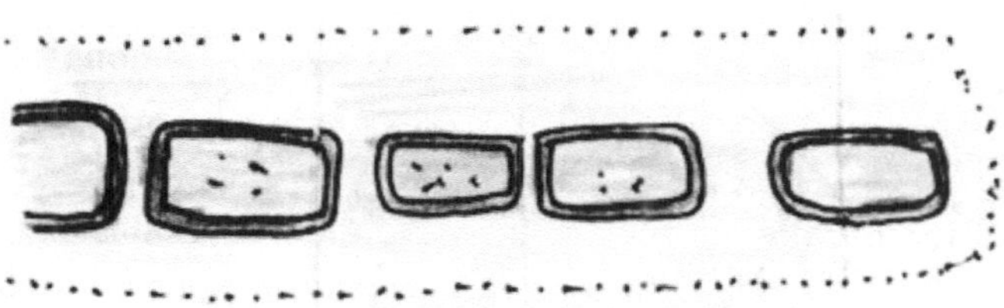

Figure 1.33: *Geminella interrupta.*

☆ In *Geminella minor* cells are close for the ends to touch.

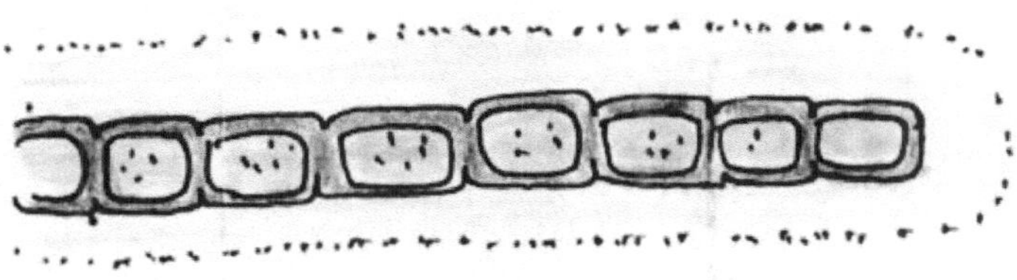

Figure 1.34: *Geminella minor.*

- Blackman considered these forms primitive that *Ulothrix* in which the cells in the filament are attached by a cementing material. Smith (1955) Supported this hypothesis.

Suggestion by Gupta and Nair (1962)

- They suggested that evolution of filament in green algae from a colonial form *Gloetoenium loilesber-gerianum* of family Oocystaceae order Chlorococcales.
- The mature thallus consists of a colony of 2, 4 or 8 cells. Four celled colonies with the cells arranged in a tetrahedral manner are of frequent occurrence.
- The cells have a tendency to get arranged in a row. Followed by repeated cell division in the same plane resulting in filamentous colonies, apparently resembling young filaments of *Geminalla* in order Ulotrichales.

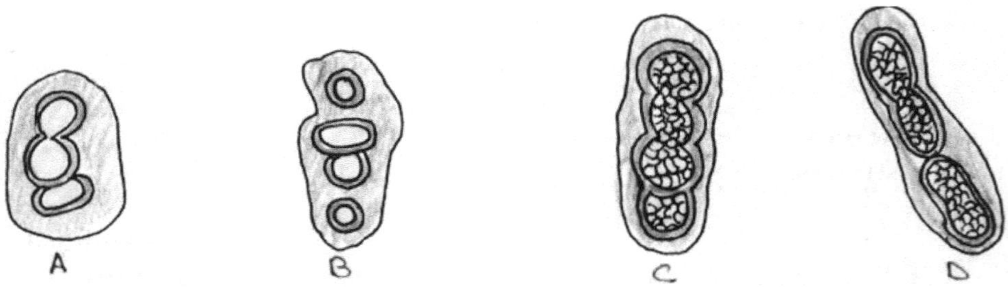

Figure 1.35

Four celled colonies of *Gloetoenium loilesber-gerianum*

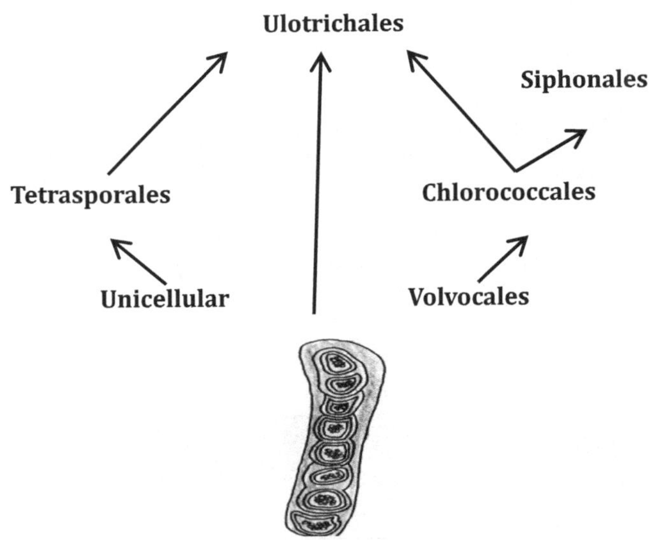

Figure 1.36: *Gloetoenium loilesbergerianum.*

Scheme of origin of filamentous habit in Ulotrichales by Gupta and Nair.

Evolution Towards Land Habit

- ☆ Land plants have advanced over the members of the chlorophycophyta along the line of these adaptations.
- ☆ Increased structural complexity of the plant body. Highly evolved reproductive organs with special adaptation to protect the gametes.
- ☆ Protection of zygote, embryo formation, and elaborate development of post-fertilization changes.
- ☆ Terrestrial algae like *Trentepholia* which grows on damp soil, bark of trees, the cell wall is stratified and the cells develop haematochrome to serve the purpose of light screen.
- ☆ Delicate apical cell protected by pecrose cap. Gametangia arise from base of Sporangia from upright threads.

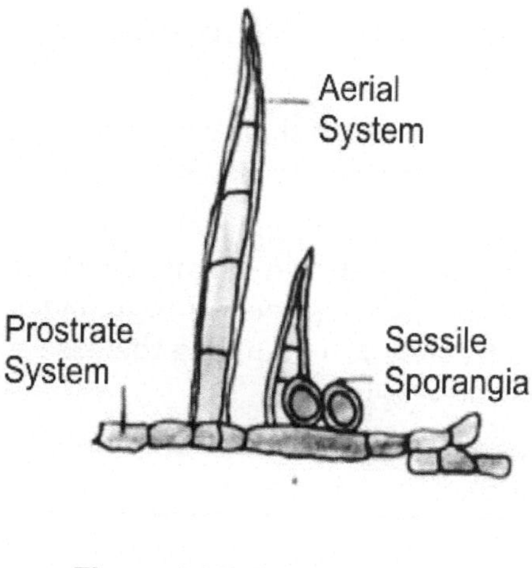

Figure 1.37: *Trentepholia* sp.

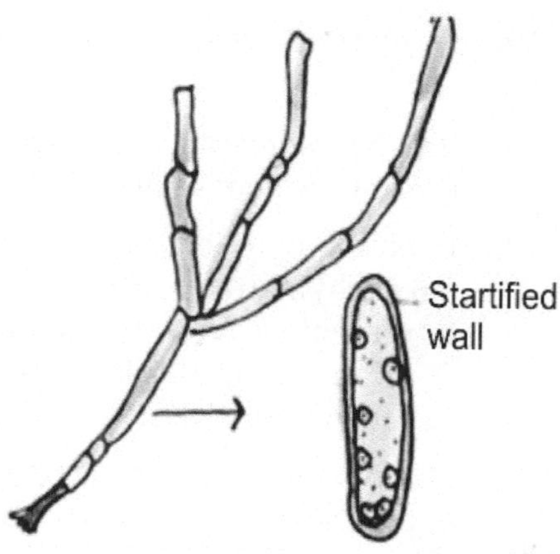

Figure 1.38: *Cladophora* sp.

- ☆ *Cladophora calcicola* which grows on rocks develops cuticle on the cell wall. Similar adaptations may be seen in different species of *Bormidium, Zygnema, Characium, Pleurococcus, Oliveria, Fritschiella.*
- ☆ Order *Chaetophorales* and *Charales* exhibit considerable morphological differentiation and show a wide range of construction from heterotrichous to highly complex aerial portion.
- ☆ Both prostate and projecting system are well developed in *Stigeoclonium.*
- ☆ Further advancement is seen in *Draparnaldiopsis* in which aerial portion in composed of long and short cells, bearing laterals of limited and unlimited growth, commonly in whorls resembling nodes and internodes.

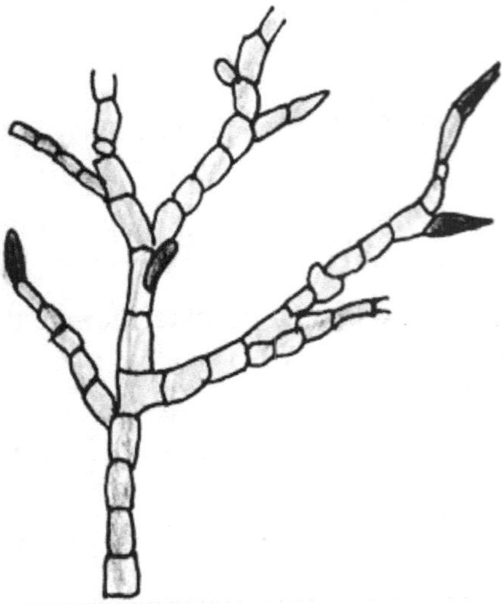

Figure 1.39: ***Stigeochlonium*** **sp.**

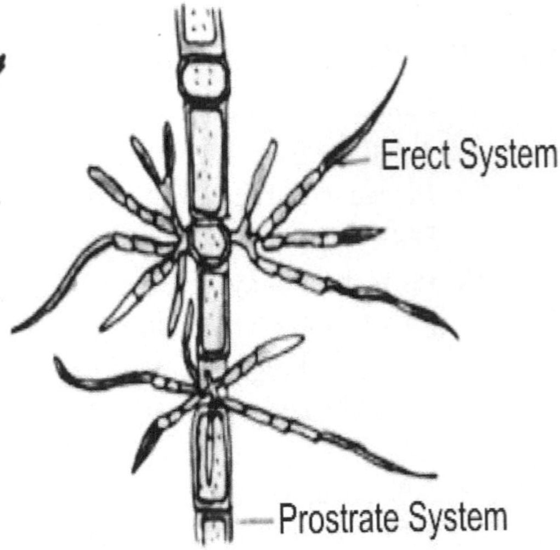

Figure 1.40: ***Draparnaldiopsis*** **sp.**

- In *Draparnaldia* aerial portion divided into long and short branches while prostate system sometimes depicted by multicellular rhizoids.
- In all the more extreme forms, the base shows a tendency to become parenchymatous, as in *Coleochaete*. Reproductive organs confined to upright system. Oogonium with trichogyne as seen in bryophytes. Development of sterile filamentous structures surrounding the base of the oogonium, that is adaptation to protect zygote.
- May be considered as a starting point of gradual elaborate development of post fertilization changes towards further complexity.

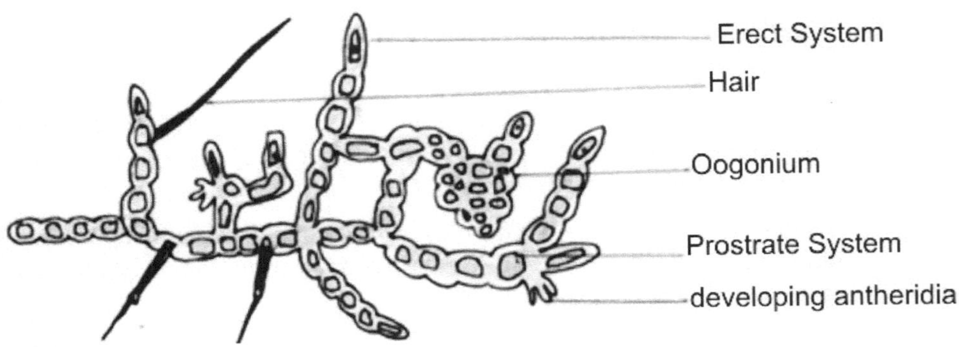

Figure 1.41: ***Coleochate*** **sp.**

- ☆ The alga *Fritschiella tuberosa* is an extremely elaborate specialized heterotrichous habit, having a rhizoidal system comparable to the rhizoids of bryophytes, a prostrate system which can grow parenchymatous structure from which arise projecting short called primary and long celled secondary systems. Rhizoids serves for anchorage and absorption, prostrate system serves for storage and reproduction, projecting system is photosynthetic.

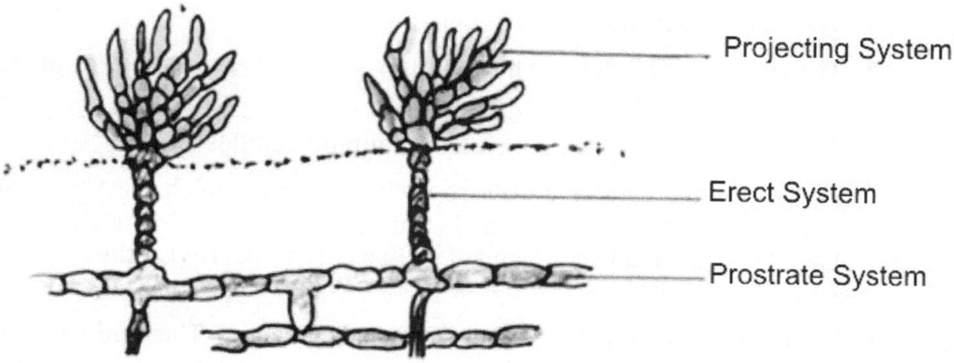

Figure 1.42: ***Fritschiella tuberosa.***

- ☆ Maximum complexity in structure of aerial portion is found in *Chara* where there is definite differentiation of nodes and the internodes being corticated. Growth in by an apical cell with a definite mechanism of protection.
- ☆ Reproductive organs are borne at the nodes.
- ☆ Prostrate system represented by rhizoids, with oblique septa resembling bryophytes.
- ☆ Reproductive organs invested with a cellular envelope, which protects the gametes against unfavourable external conditions.
- ☆ Oogonium well protected by sterile covering which can serve the purpose of protection of zygote.
- ☆ Elaborate post fertilization changes and embryo development is indicated.

Conclusion

Though many green algae represent 'dead ends' as far as subsequent evolutionary development is concerned, there is no doubt that they are extremely successful forms, since they are able to exist in a wide range of habitat all over the earth. The great diversity among the green alga would suggest that many forms among these plants would be able to adapt themselves to the changing environmental conditions.

Points to Remember

The Chlorophyta are characterised by the following:

1. They are grass green in colour owing to the preponderance of chlorophyll *a* and *b* over carotene and xanthophyll.
2. The pigments are localised in the green plastids known as chloroplasts.
3. The reserve carbohydrate food is stored as starch.
4. The chloroplasts normally contain the pyrenoids.
5. The cell has a well-defined nucleus and in the higher forms a central sap cavity in addition.
6. The cell wall is stable and invariably contains cellulose.
7. The majority produce motile reproductive cells which may be bi-or quadriflagellate rarely with a ring of fiagella as in Oedogoniales. The flagella are of equal length and of whiplash type inserted at the anterior end.
8. Sexual reproduction ranges from isogamy to oogamy. The zygote nearly always enters upon the resting period.
9. The sex organs are always unicellular.
10. Zygote generally is the only diploid structure in the life cycle.
11. The plant body is unicellular or multicellular. The former may be solitary or colonial. The multicellular form show a considerable range of variation in the form and structure of the plant body. Commonly they are filamentous, sometimes massive. However it is always simple in construction and shows relatively little cellular organisation.

Self Learning Materials (Chlorophyceae)

(Green Algae)

1. Introduction

- ☆ The division Chlorophyta more appropriately called Chlorophycophyta. This group comprising of about 20,000 species. These are included in a single class Chlorophyceae.
- ☆ It is evident tha the cell structure of a grean alga resembles that of the higher plants in the following essential details.
- ☆ Presence of Chlorophyll-a and Chlorophyll-b as the principal pigments.
- ☆ The accessary pigments Xanthophylls and Carotenes are the same in both.
- ☆ The principal photosynthetic food reserve in starch.
- ☆ The location of Chlorophyll in the Chloroplasts.

- ☆ The components of cell such as the cytoplasm, nucleus, cell membrane and chloroplast have basically the same structure as those of the higher plants.

2. Objective

Three different hypotheses, which attempt to explain this debatable point have been put forth.

- ☆ The first suggests an origin of multicellular condition in green algae from the motile unicellular ancestor.
- ☆ The second from the origin of Filamentous (multicellular) palmelloid forms (Tetrasporales).
- ☆ The third envisages origin from the colonial coccoid form (Chlorococcales).
- ☆ Most of the algologists agree that both the Tetasporales and Chlorococcales have directly evolved from the motile green unicellular ancestor like Chlamydomonas.

3. Content

Morphologically it may consist of motile or non-motile unicells – Motile and non-motile colonies simple and branched filaments – Siphonaceous and Coenocytic filaments – foliaceous thallus – Cushion like thalli and highly evolved heterotrichous filament. However its always of a simple construction, it shows no differentiation into true root, stem and leaves for this reason the plant body of the algae is called a thallus.

4. Revision Points

- ☆ Habit: Unicellular, colonial, coenobial, filamentous Siphonaceues and Thalloid forms.
- ☆ Cells Structure: Eukaryotic, well organised nucleus and other cell organelles present.
- ☆ Cell Wall: It consist of two concentric layers. The inner layer in cellulose in nature. The outer is a pectose layer covered with mucilage sheath.
- ☆ Flagella: Motile cells with 2, 4 or more Flagella: Flagella anteriorly inserted equal in length and whiplash type.
- ☆ Distribution of pigments: Chlorophyll a and b α- and β-carotene, Xanthophyll.
- ☆ Reserve food material: Starch and fats.
- ☆ Reproduction: For chlorophyceae reproduction by vegetative, a sexual and sexual means: sexual reproduction by isogamous, aniso gameous or oogamous.

5. Intext Question

1. Write an account of origin of green algae.
2. What is Coenobium? Give three examples.
3. Describe briefly range of thallus structure in Chlorophyceae.
4. Write a notes on the structure of algal flagella with suitable diagram.
5. Compare and contrast the sexual reproduction in Ulothrix, Spirogyra and Oedogonium.

6. Summary

- ☆ Diversity and distribution mostly in fresh water, a few in brackish and saline water few are terrestrial.
- ☆ It show wide range of various in their thallus structures like Unicellular motile and non-motile, Coenobium, Palmelloid, Dendroid Filamentous, branched and unbranched heterotrichous, siphonaceus and parenchymatous.
- ☆ Pigments are located in the Chloroplast which has pyrenoids.
- ☆ Reproduce by vegetative (cell division and fragmentation), a sexual (zoospore, aplanospore, akinete *etc.*), and sexual isogamy (to oogamy).
- ☆ Zyospore or oospore only diploid structure in the life cycle.

7. Terminal Exercises

1. Non-motile unicellular alga in

   a) Chlamydomonas b) Chlorella

   c) Pandorina d) Voloox

2. Cup or bell shaped chloroplast is present in

   a) Spirogyra b) Nostoc

   c) Chlorella d) Hydrodictyon

3. Describe the various types of Zoospores in Ulothrix
4. What you understand the term Isomorphic alternation of generation? Explain it with the help of life cycle of an alga studied by you.
5. Write short notes on: a) Sea Lettuce, b) Reproduction in Ulva.
6. Where does Ulva occur in India.

8. Assignments

1. Write an account of habitat, structure and reprodution of Chlorophyceae member.

2. Describe the characters of the order Ulvales. Give an account of reproduction in a member of the order.
3. Write an illustrated account of the sexual organs of Chara and Nitella and germination of its Oospores. Mention how it differs from other green algae.

9. Learning Activities

a) Describe the habit, structure and methods of reproduction in Oedogonium and indicate features of special interest in its life history.

b) How does the Macrandrous specis of Oedogonium differ from Nanandrous species in the method of sexual reproduction?

c) Give an account of the occurrence, structure and reproduction in Chlorella, briefly describe its economic importance.

10. Keywords

Grass green in colour, Cell wall – cellulose, Reserve carobohydrate food in stored as starch, Sex organs are always unicellular, Isogamy to oogamy through anisogamy.

CHROMOPHYTA

XANTHOPHYCEAE

General Characteristic Feature of Xanthophyceae

1. Cell wall is made up of two layers, the main components of which are pectic compounds
2. Important pigments are chlorophyll a, chlorophyill e,B-carotene and xanthophylls like lutein, violaxanthin and neoxanthin.
3. Reserve food materials are fat or oils, starch is absent and pyrenoids are very rarely present.
4. Motile bodies contain two unequal flagella
5. Chromatophores are discoid in shape.

Classification

→ Heterococcales-coccoid forms

→ Heterochloridales-Motile forms

Xanthophyceae →

→ Heterotrichales-Filamentous forms

→ Heterosiphonales-siphonaceous forms

VAUCHERIA: De comdole

Systematic Position

Class	: Xanthophyceae
Order	: Hetrosiphonales
Faimly	: Vaucheriaceae
Genus	: Vaucheria.

General Features

1. It is commonly found in fresh water ponds or on moist soil in the form of green mat.
2. The Plant body consists of a well branched siphon like, coenocytic filaments *i.e.* the filamentous are aseptate, multinucleate continuous structures.
3. Terrestrial species are attached to substratum by means of a branched hyaline cell known as hold fast or hapteron.
4. The cell wall is very thin, consisting of outer pectose and inner cellulose layers.
5. Reserve food material is present in the form of oil drops, lipid and leucosin or chrysolaminarin pyrenoids are absent.
6. The nuclei are present in the cytoplasm towards the central vacuole.

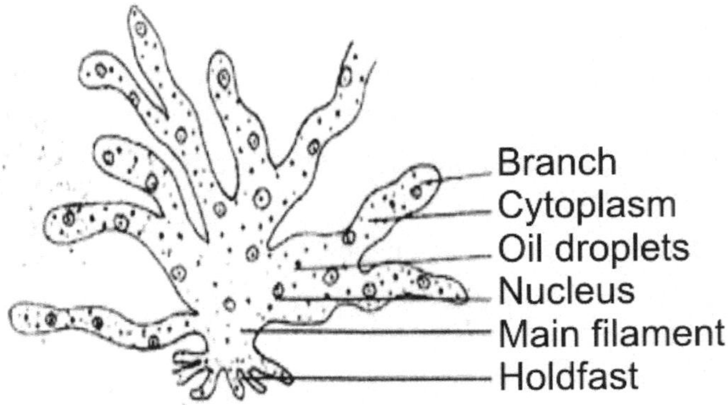

Figure 1.43

BOTRYDIUM: Wallroth, 1815 (a small cluster)

Class	: Xanthophyceae
Order	: Heterosiphonales
Faimly	: Botrydiaceae
Genus	: Botrydium

External Features

1. It is a unicellular, coenocytic, terrestrial alga consisting of an aerial and an underground portion.
2. The aerial portion is spherical or pear shaped vesicle, measuring 1-2mm, in diameter.
3. Plant body is differentiated into an underground rhizoidal portion and an aerial or sub aerial vesicular portion.
4. Vesicle is spherical or balloon shaped and green in color but in the shady conditions the visicle becomes elongated.

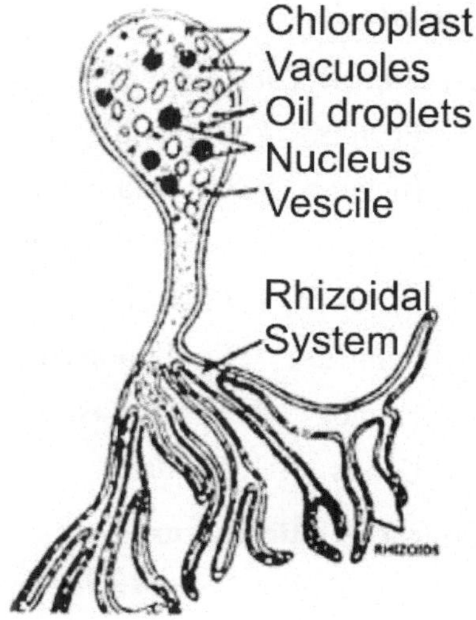

Figure 1.44

5. Rhizoidal Portion is well branched, underground, colourless and aseptate.
6. Pyrenoids are present when the vesicles are young but whether these bodies are present in the mature condition also –is not clear.
7. Reserve food material is in the form of oil or leucosin starch is totally absent.

BACILLARIOPHYCEAE

Important characteristic features are:

1. Widely distributed in sea, freshwater and soil.
2. Chromatophores are yellow or golden yellow in color.
3. Reserve food material is in the form of oil or volutin granules.
4. Pyrenoids are present but very lack starch sheath.
5. All the members are unicellular but form colonies when many members come together.
6. Cell wall is made up of pectic material and silica.
7. The cell wall consists of two halves arranged in the manner of a box and lid. The outer half is known as epitheca and the inner half as hypotheca.
8. Members are either radially symmetrical (order centrales) or bilaterally symmetrical (order pennales).
9. Centrales: Radially symmetrical, raphe absent. Many chromophores per cell, reproduction by auxospores.
10. Pennales: Symmetrical or even asymmetrical, presence of raphe, one or two chromatophores per cell, reproduction by auxospores and statospores.

Cyclotella

Systematic Position

Phylum	: Chrysophyta
Division	: Chromophyta
Class	: Bacillariophyceae
Order	: Centrales
Suborder	: Coscinodiscineae
Family	: Thalassiosiraceae
Genus	: *Cyclotella*

Characteristic Features

1. Cells solitary, valves tangentially undulate and valve wall is alveo-lated.
2. A central field distinctly different from the rest of the valve.
3. Central portion with pflexes and coarsely punctate.
4. Cells in chains fairly close, separated by occluded process on marginal part valve.

A. ***Discostella asterocostata*** B. ***Cyclotella meneghiniana***

Figure 1.45: Centric Diatom.

Pinnularia

Systematic Position

Phylum	: Chrysophyta
Division	: Chromophyta

Class : Bacillariophyceae

Order : Pennales

Family : Naviculoideae

Genus : Pinnularia

Characteristic Features

1. It is a unicellular, microscopic alga, which is commonly known as diatom.
2. Each cell is oval or elliptical in shape with bilateral symmetry.
3. Each cell consists of a frustule, which is composed of two overlapping halves (theca), the epitheca and hypotheca.
4. Epitheca overlaps the hypotheca and the overlapping portion is known as girdle or connecting bond.

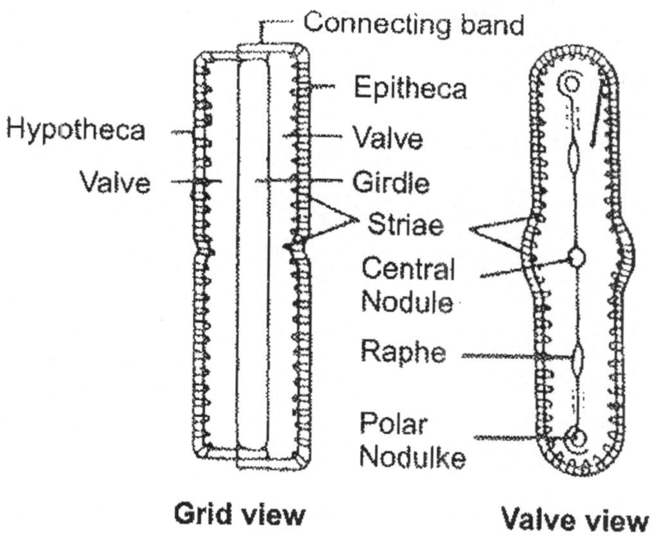

Figure 1.46

5. Striations are arranged isobilaterally, *i.e.*, on either side of the central line.
6. The central line has a longitudinal slit which runs from one polar nodule to the other and is known as raphe.
7. The raphae has two polar nodules and a central nodule. These nodules are the internal thickenings of the wall.

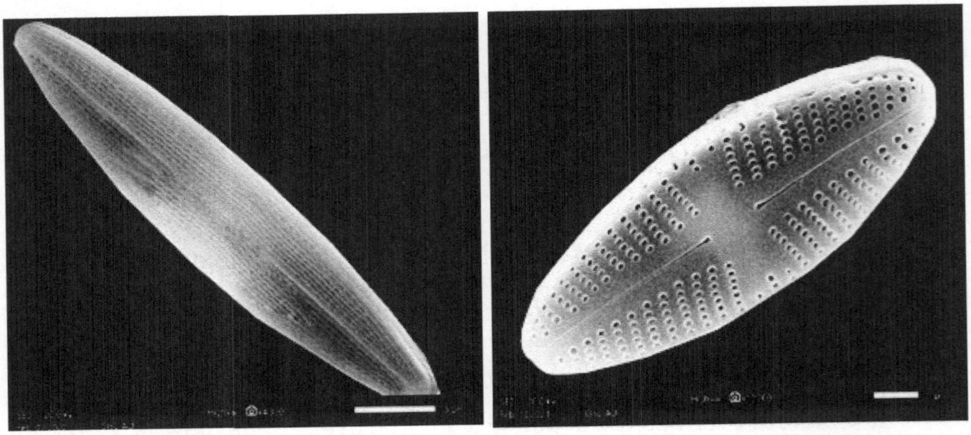

A. ***Navicula symmetrica*** **B.** ***Navicula seminuloides***

Figure 1.47: Pennate Diatom

Unit 2

Phaeophyceae

Important characteristic features are:

1. Pigments are chlorophyll a and c, β-carotene, flavoxanthin, fucoxanthin, violaxanthin and luetin.
2. Due to the presence of fucoxanthin, the members are brown in color and are commonly called brown algae.
3. Reserve food materials are mannitol and laminarian. Pyrenoids are present only in some lower forms.
4. Reproduction units are motile having two flagella attached laterally.
5. Motile reproductive bodies are developed either in unilocular or Plusriliocular sporangia
6. Sexual reproduction ranges from simple isogamy to oogamy.

The class includes nine orders

1. Ectocarpales
2. Tilopteridales
3. Cutleriales
4. Sporochanales
5. Desmarestiales
6. Laminariales
7. Sphacelariales

8. Dictyotales
9. Fucales

Anatomical Features

1. The outermost covering layer on both side in a single called thick meristoderm or epidermis consists of small rectangular cells. They are uninucleate, full of chromatophores and help in photosynthesis.
2. Meristoderm is followed by central medulla which is composed of large rectangular cells with few or no chromatophores. These cells are meant for storage of reserve food and contain fucosan vesicles and reractive granules, *etc.*
3. Multicellular hair in the form of the tufts may be observed on both the surface of thallus.
4. In between meristoderm and medulla few layers of cortex may be perfect sometimes medulla is many called thick.
5. Asexual reproduction takes place by means of tetrapors, sexual reproduction is oogamous and the plants are strictly dioecious.

Ectocarpus

Systematic Position

Division : Phaeophyta
Class : Phaeophyceae
Sub Class : Isogeneratae
Order : Ectocarpales
Family : Ectocarpaceae
Genus : *Ectocarpus*

Characteristic Features

1. This marine alga is the most primitive of all the brown algae and comprises many species and a few have been reported to occur in freshwater.
2. The thallus is fine, tufted, brownish heterotrichous filament consisting of well defined erect and prostrate system.
3. The filaments are generally uniseriate and usually consist of a single stand of cells.
4. The branching is always lateral, end in colourless mucilage hairs.
5. The cells are short and cylindrical; cell wall is made up of pectic-cellulose. The protoplasm consists of golden brown disc or band shaped chromatophores with characteristic pigment fucoxanthin.

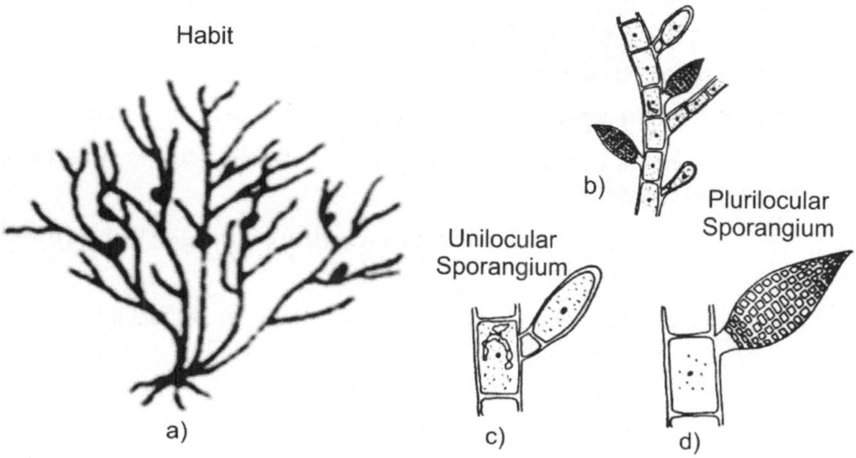

Figure 2.1

6. Genetically the thalli of Ectocarpus are of two kinds, haploid and diploid. Morphologically they are alike.
7. The haploid thalli are concerned with sexual reproduction and diploid with asexual reproduction.
8. The diploid plants bear two kinds of sporangia
   i. Unilocular-produce haploid zooids or swarmers called meiozoospores
   ii. Plurilocular-Gives rise to single diploid zoospore
9. Sexual reproduction is isogamous, take place by the formation of biflagellate gametes structurally similar to the zoospores produced in gametangia.
10. The haploid and diploid plants are morphologically alike. Ectocarpus thus exhibits an isomorphic type of alternation of generations.

Padina

Systematic Position

Division : Phaeophyta

Class : Phaeophyceae

Order : Dictyotales

Family : Dictyotaceae

Genus : *Padina*

Characteristic Features

1. *Padina* is a marine algae. The thallus consists of stalk and fan shaped fronds, 5-12 cm in height, the larger ones often loosely rolled on their longitudinal axis like a cornet are distinctive of all the species of the genus.
2. The stalk is the upward continuation of the prostrate rhizome.
3. The rhizome is very much branched and attached to the substratum by tufts of rhizoids.
4. The thallus shows concentric zones which are arranged in 4-8 rows. They are formed by the hairs.
5. The zones of hairs occur alternately on the two surfaces but are more strongly developed on the upper and are stated to be specially prominent in plant exposed to strong light.
6. The thallus shows an apical cell and row of marginal initials. The older portion of the thallus has the maximum of 10 or more cells in thickness.
7. The cells are more are less isodiametric and with iodine.
8. The longitudinal section of the thallus from the apex to the interior shows circinate coiled vernation.
9. These cells are in the protection of the apical cells.
10. On the thallus surface, trichomes and sporangia or gametangia of two kinds are found on the older parts. The gamete producing organs are known as gametangial sours.

a) *Padina* Habit

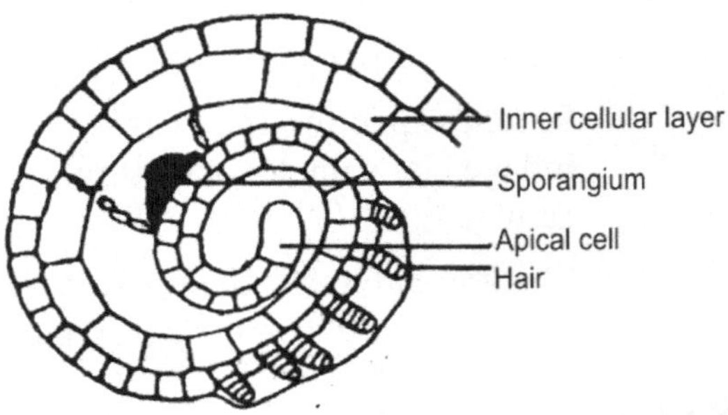

b) *Padina* Thallus L.S.

Figure 2.2

Dictyota: Lamouroux, 1809

Class	: Pheophyceae
Order	: Dictyotaceae
Genus	: Dictyota

General Characteristics

1. Well developed and branched Thallus is attached to the substratum by a disc like hold fast
2. Hold fast is prostrate and irregular in shape
3. From the hold fast arise many erect, branched leafy structures or fronds.
4. A frond consists of a lower cylindrical stalk and an upper well branched portion.
5. Fronds are flattened structures and lack a midrib
6. At the apex of each branch is present an apical cell.
7. Apical cell divides into two, thus developing into two branches of dichotomy.

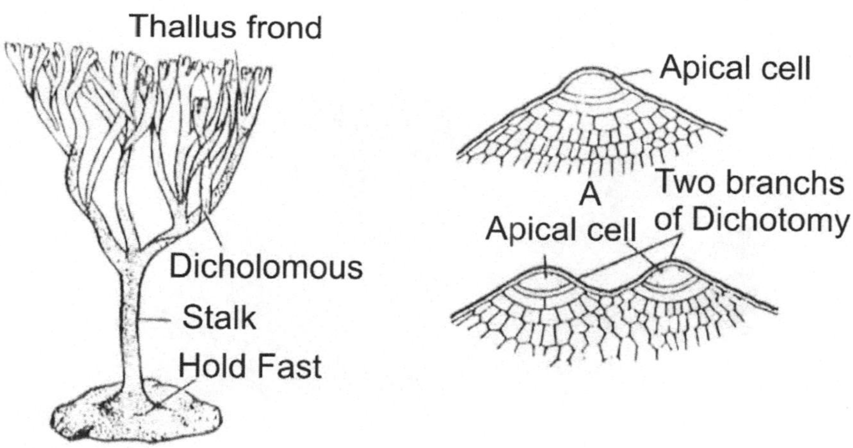

Figure 2.3: *Dictyota dichotoma* Thallus.
Upper part of the frond is dichotomously branched.

Stoechospermum

Systematic Position

Division	: Phaeophyta
Class	: Phaeophyceae
Order	: Dictyotales
Family	: Dictyotaceae
Genus	: *Stoechospermum*

Characteristic Features

1. Thallus is large tufts, 15-25 cm or more high.

2. Base with irregularly ramified, bent, decumbent branches interwoven
3. Rhizoids numerous from the base, attaching the plant to the substratum.
4. Thallus erect, spathulate, dichotomously branched repeatedly a few times, flat, without midrib; segments of thallus 1-2 cm broad; margin entire; apex bifid or flatly tuncate.
5. Section of thallus shows greater part with large parenchymatous cells in the middle and on either side covered by two layers of small cells.
6. Hairs in groups scattered on both sides of thallus.
7. Tetrasporangia in longitudinal rows in irregular outline near the margin and extending to some distance along the margin of the thallus, 0.2-0.3 cm thick.
8. Thallus brown in younger parts, tending to be dark brown in the older and basal part of the thallus.

Figure 2.4

Sargassum

Systematic Position

Division	: Phaeophyta (Phaeophycophyta)
Class	: Phaeophyceae
Order	: Fucales
Family	: Sargassaceae
Genus	: *Sargassum*

Characteristic Features

1. *Sargassum* is a marine alga popularly called the gulf weed, occurs in tropical or temperate zones attached along the rocky shores.
2. It is macroscopic and more a less bushy in habit, looks like a small angiospermic plant found attached to the substratum by means of well developed hold fast.
3. The thallus is mainly composed of the much branched primary, long laterals.
4. The lower members of the auxillary branch system may be reduced to air-bladders. The subsequent ones are fertile and may be cylindrical to flattened called receptacles.
5. It reproduces entirely by fragmentation and sexual reproduction.
6. Sexual reproduction is oogamous. The sex organs are lodged in conceptacles which are confined to the receptacles.
7. Some species are monoecious and others Dioecious.
8. The zygote germinates without undergoing any resting period and forms the diploid thallus.
9. *Sargassum* is a source of alginates and alginic acid.

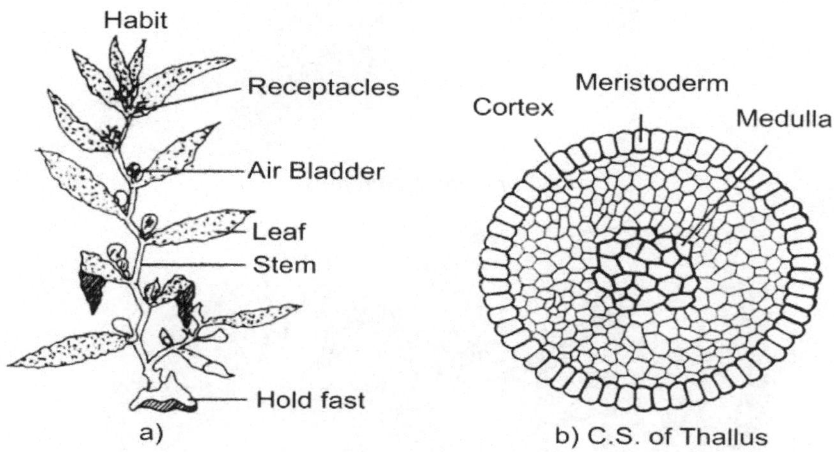

Figure 2.5

Turbinaria

Systematic Position

Division : Phaeophyta

Class : Phaeophyceae

Order : Fucales

Family : Sargassaceae

Genus : *Turbinaria*

Characteristic Features

1. *Turbinaria* is a multicellular marine algae.
2. It is found round the year on the rocks, stones and old corals in the sub tidal region. It occurs only in less quantity.
3. The thallus in yellowish brown or brown in colour 20-25 cm in height, possess elongated cylindrical branches, leaves possessing angular margins.
4. It is used as raw material for the manufacture of sodium alginate and also yields 8-10 per cent of mannitol.

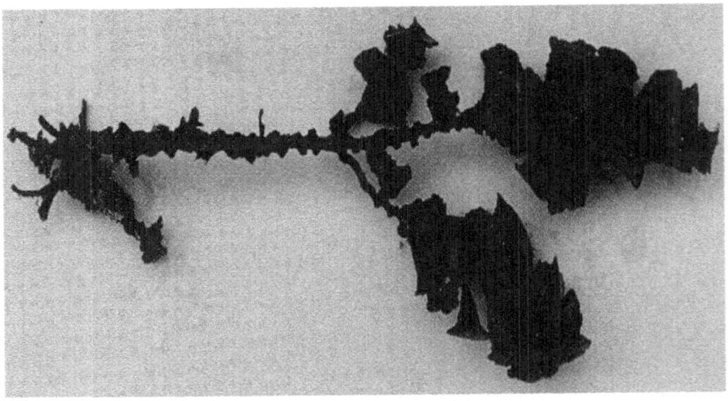

a) *Turbinaria conoides*

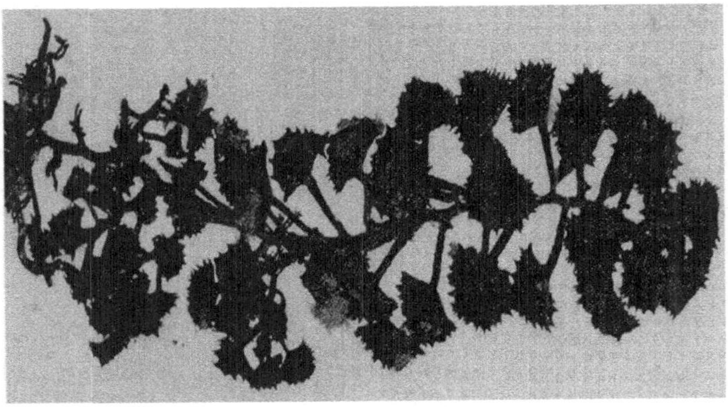

b) *Turbinaria* sps.

Figure 2.6

Habit. The *Sargassum* plant is macroscopic and more or less bushy in habit. Apparently it looks like a small angiospermic plant. The plant body is radially symmetrical (5". *patens*). Some species such as *S. vulgare* and *S. filipendula* are found attached to the substratum. Free floating unattached forms like *S. natans* and *S. baciferum* are also fairly common. *S. natans* occurs in huge floating masses. It has no attaching disc or holdfast and is sterile. It multiplies exclusively by fragmentation of the thallus. The other species *of Sargassum* found in the Sargasso sea is *S. hystrix.*

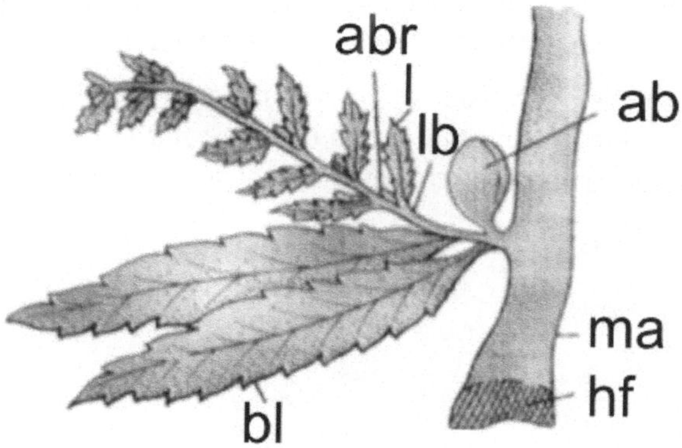

Figure 2.7: Basal Portion of the Plant of *Sargassum enerve.* *ab:* Air bladder; *abr:* Axillary branch; *bl:* Basal lateral; *lb*: Lateral branch; *ma:* Main axis; *hf:* Holdfast.

Organisation of the Thallus

(Figure 2.7). *Sargassum* plant is a diploid sporophyte. The main or central axis (*ma*) in some species (*S. filipendula*), is 30 cm. or more in length. It bears richly branched primary laterals. The phyllotaxy is 2/5. Each primary lateral or branch (*lb*) is of unlimited growth. It is also called the long shoot. The long shoot bears numerous secondary laterals of limited growth. Of the latter the basal one (*bl*) is usually the most conspicuous. It is shaped like a leaf. It subtends more or less clearly the remainder of the branch system. These so-called leaves are in some cases very narrow and drawn out with a serrate margin. In *S. patens* these leaf-like expansions have a conspicuous mid-rib. The mid-rib is absent in the secondary laterals of *S. enerve* (*bl*). The secondary laterals are flattened in a vertical plane. In the lower part of long shoots these so-called lower leaves or laterals of limited growth bear axillary rudiments (*abr*). In the upper region of the adult plant they subtend apparently considered axillary branch systems. The basal member or members of axillary branch system are swollen into stalked and rounded structures called the air bladders. The subsequent members of the axillary branch system may be cylindrical or flattened. They are called the receptacles. The receptacles are studded with fertile, flask-shaped conceptacles. The receptacles are produced when the plants become fertile. In *Sargassum* the receptacles are specialised branches and not metamorphosed vegetative branches. In some species the so-called leaf subtending the branch system may be modified into an air bladder. Such air bladders are homologous to the laterals of limited growth of "leaves". They may (*S. longifolium*) or may not be terminated by the leafy growth. The air bladders contain the same gases as the atmosphere. Hence they act as floats keeping the plant upright in water. In the unattached species they help the plant to float. Colla and Willis believe them to be respiratory in function while others look upon them as a mechanism for decreasing specific gravity. They arise as surface outgrowths accompanied by increase in the thickness of the cortex.

Sargassum plant is anchored to the substratum by means of an attaching disc. It has an irregular outline and more or less rough appearance. In addition numerous stolon-like structures arise from the base of the main axis. They provide additional anchorage.

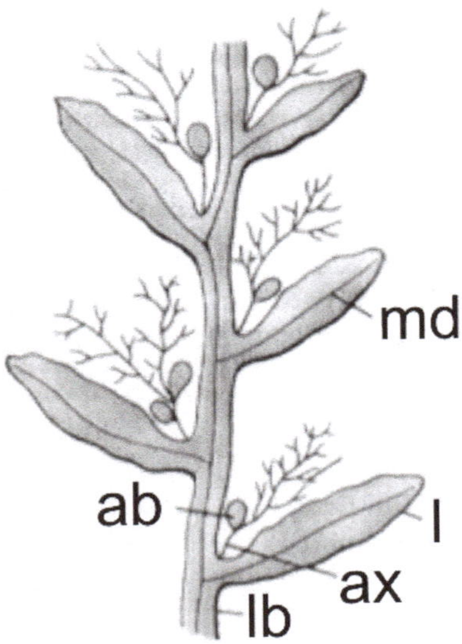

Figure 2.8: *Sargassum*.
***lb*: Lateral branch; *l*: Bearing lateral; *ax*: With axillary branches; *ab*: Air bladder; *md*: Midrib; *re*: Receptacles bearing conceptacles.**

S. enerve turns green when dried. Consequently it is often used as a piece of decoration. In *S. longifolium,* which is a South African species, the main axis is flattened. The laterals arising from it show distichous arrangement (Figure 2.8). Each lateral bears two basal leaves alternately arranged. The upper one is modified into a bladder surmounted by a leafy growth. The branch axis beyond the two basal leaves bears a series of branched receptacles.

Structure of the Thallus (Figure 2.9, A)

Viewed in cross section the thallus reveals the following structure:

1. *Meristoderm.* It is a peripheral layer of columnar cells without any air spaces between them. The cells of this layer contain abundant plastids and the fucosan granules. This palisade layer (*p*) of cells constitutes the *meristoderm.* It is concerned with photosynthesis. On the outside the palisade cells are covered by a mucilaginous cuticle. On the inner side of the palisade layer is a hypodermal layer (*hy*) consisting of similar cells. The cells of both the layers have pits in their walls (Figure 2.9 B).
2. *Cortex.* Within the hypodermal layer is the cortex (cort) consisting of thick-walled polygonal cells. The cortex may be several layers thick. It probably functions as the storage region and plays a mechanical role. The innermost cortical cells have their walls gelatinised and merge imperfectly into the central zone.
3. *Medulla.* The central zone of the thallus is called the medulla (M). The cells in this region are loosely arranged. They are narrow and elongated. The outer cells of the medulla have thick walls whereas the inner have thin walls. There is, however, no gelatinized material. Hanstein described scalariform thickenings on the lateral walls of the medullary cells. The medullary hyphae are absent. In the so-called leaves the medulla is composed of thin-walled cells only. The medullary region is almost absent in free floating species. The attaching organs contain hyphae in the medullary region. Medulla probably serves as a food conducting region. According to Hanstein the cells of all the zones are inter-connected by pores in their cross walls. Others, however, deny this.

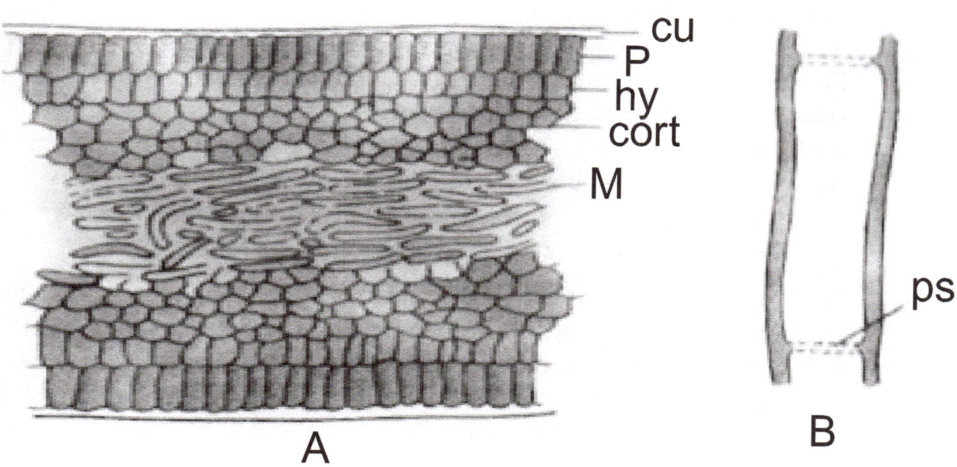

Figure 2.9 (A-B). *Sargassumsp.* (A), T.S. Thallus (*cu,* cuticle; *cort,* cortext; *hy,* hypodermis; M, medulla; ps, palisade layer);
(B), Single cortical cell showing perforated transverse septa (*ps*).

The anatomy of the thallus of *Sargassum* reveals that the vegetative cells constituting it show a great degree of differentiation accompanied by division of labour. It contains tissues specialised for photosynthesis, storage and support, conduction and attachment.

Reproduction

It takes place by vegetative and sexual methods only. Asexual reproduction by the formation of spores is absent.

1. *Vegetative Reproduction. Sargassum* reproduces vegetatively by fragmentation. The older parts die and decay separating the younger parts. The latter continue to grow and finally develop into full-fledged *Sargassum* plant. *S. Natans* and *S. hystrix* multiply exclusively by this method.
2. *Sexual Reproduction.* It is the usual method of reproduction and is **oogamous.** The sex organs are produced in flask-shaped cavities called the **conceptacles** borne on the sporophyte plant.

The conceptacles are aggregated in the upper region of the axillary branch systems. These may be cylindrical or flattened and are called the receptacles.

(*a*) *Mature conceptacle.* It is a depression or cavity (Figure 2.10B) in the thallus. It lies in the superficial part of medulla of the receptacle. The conceptacle has a wall of its own. The wall consists of a few layers of compactly arranged flat cells rich in chromatophores.

The conceptacle wall separates the cavity from the loose tissue of the medulla. The greater part of the wall bears many branched hair-like filaments, the paraphyses. The latter consist of barrel-shaped cells with scanty chromatophores. The septa between the cells are often oblique. Each conceeptacle opens to the exterior by means of an aperture termed the ostiole. Near the aperture are present the colourless hair. They are called the periphyses. Between the paraphyses and the colourless hair near the ostiole are present hairs with abundant chromatophores.

(*b*) *Development of conceptacle* (Figure 2.10, *A-G*) It is formed from a single superficial, flask-shaped initial cell (*A, ic*). It is distinguishable from others by its larger size, slower rate of division and large nucleus. The conceptacle initial divides by a curved septum into two unequal daughter cells (C). The upper daughter cell is called the tongue cell (C, *tc*) and the lower basal cell (C, *be*). In the meanwhile the cells surrounding the conceptacle initial divide in such a manner that it sinks down. The tongue cell remains separate from the adjoining tissue. The basal cell divides repeatedly by vertical walls to form the wall of the conceptacle except its upper narrow portion. The latter is formed from the adjoining tissues. The tongue cell merely elongates and in some cases divides to form a row of few cells (*G, tc*). Eventually the tongue cells disappear.

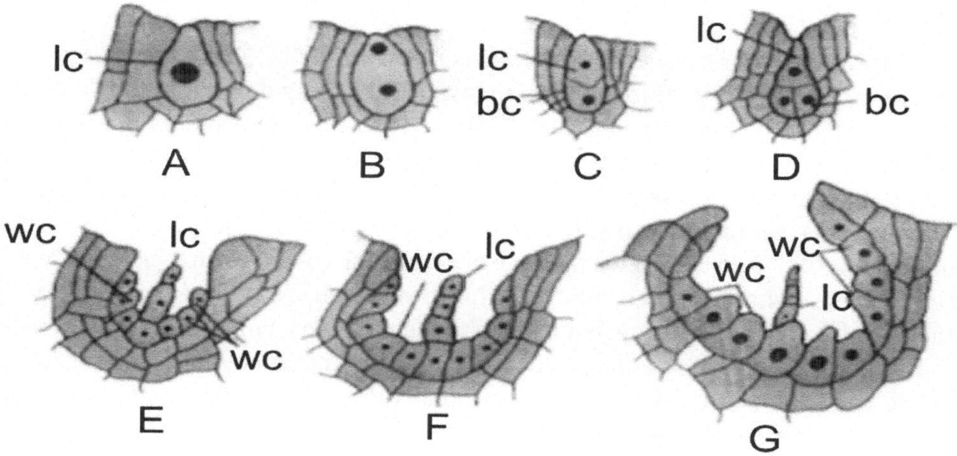

Figure 2.10: (A-G). Development of Conceptacle of *Sargassum* sp. (A-D), Earlier stages. (A), Flask shaped initial (*ic*); (B), Nuclear division; (C), Formation of basal cell (*b c*) and tongue cell (*tc*); (D), Basal cell divided into two by a vertical wall; (H-G). Later stages in development of the conceptacle.

Sex Organs

The distribution of sex organs varies in different species. Some species are monoecious and others dioecious. The conceptacles are always unisexual. In the monoecious species the male and the female sex organs are developed in separate conceptacles borne on the same plant. In the dioecious species the male and female conceptacles are borne on different plants. The dioecious plants are thus either male or female. The male and female plants are to some extent distinguishable. The male receptacles are smooth and the female spinous. In *S. horneri* the male receptacles are called the antheridium and the female oogonium.

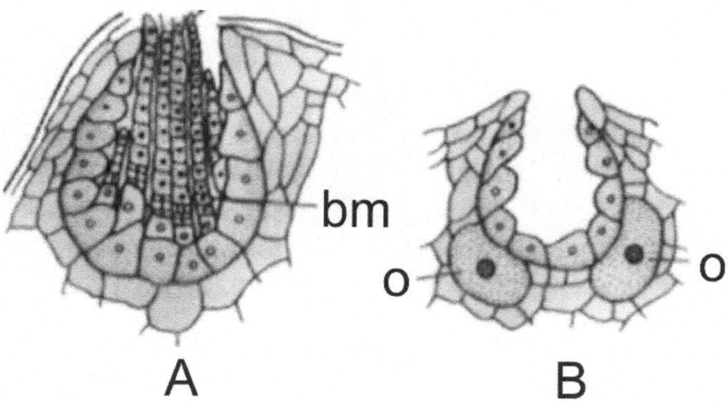

Figure 2.11 (A-B). (A), Vertical section of young cryptoblast. bm_t basal meristem; *h,* hair; (B), VS. of young female conceptacle with two embedded oogonia (*o*).

Oogonium (Figure 2.11, B)

The oogonia are sessile and borne directly on the wall of the conceptacle. They are, as a matter of fact, embedded in the wall. The young oogonium is filled with dense cytoplasm. It has a single diploid nucleus and contains plenty of oil drops. During maturation the diploid nucleus undergoes three successive divisions. The first two constitute **meiosis.** As a result, eight haploid daughter nuclei are formed. The cytoplasm undergoes no cleavage. Towards maturity the protoplast of the oogonium rounds off to form a single egg or ovum. The young ovum is eight nucleate. Out of these eight nuclei one enlarges. The other seven are, at first arranged along the periphery. They finally degenerate during fertilization. Some investigators hold that the nuclear divisions take place when the oogonium is young and enclosed in the conceptacle. Rao and others hold that the nuclear divisions occur when the oogonia are discharged from the conceptacle and remain outside the ostiole attached to the conceptacle wall. At maturity the oogonial wall thickens and gets differentiated into three layers. These are the outer thin exochite, the middle thick mesochite and the innermost endochite.

The mature oogonia (Figure 2.12) are discharged through the ostiole. Even after emergence each oogonium remains attached to the conceptacle wall by means of a long gelatinous stalk (*mst*). The oogonial stalk is formed by the elongation of the thickened apex of the oogonium. The distal end of the thickened apex elongates and curves back till it fixes itself to the wall of the conceptacle. The curved stalk then straightens out. Consequently the entire oogonium is pushed out through the ostiole in an inverted position. The other viewpoint is that the exochite ruptures. The mesochite forms the long mucilaginous stalk which serves to attach the discharged oogonium invested by the endochite.

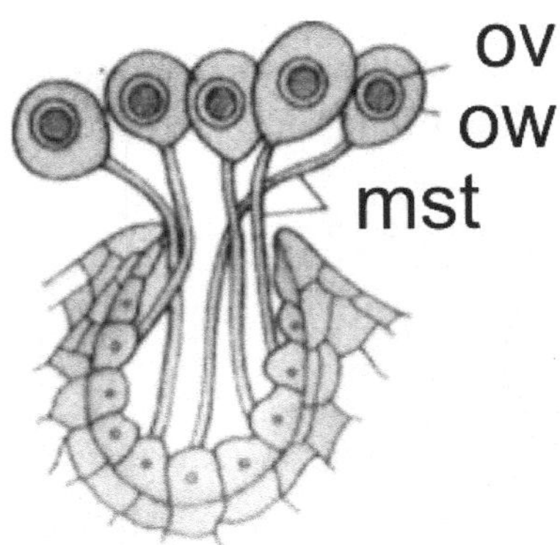

Figure 2.12. ***Sargassum*** **sp. Extruded oogonia attached to the conceptacle wall by long gelatinous stalks (*mst*); *ov*, ovum; *ow*, oogonial wall.**

Development of Oogonium (Fig 2.13, A-F). Any cell in the wall of the conceptacle may function as the oogonial initial (A, *oi*). The latter grows in size (B, *oc*). It then divides by an asymmetrical transverse division to form two unequal cells (C). Of these the lower smaller one is called the stalk cell (*sc*) and the upper oogonial cell (*oc*). The stalk cell undergoes no division. This accounts for the embedded nature of the oogonium. The oogonial cell increases in size. It is filled with dense cytoplasm and is devoid of vacuoles. It has plenty of oil drops. Its single nucleus undergoes three successive divisions to form eight daughter nuclei (D). Of these the first and the second division follow rapidly one upon the other and the third after a short period of rest. The first two nuclear divisions constitute meiosis. In *Sargassum,* reduction division thus takes place during gamete formation. The nuclear division in the oogonium is not followed by cleavage of the cytoplasm, the protoplast of the oogonium withdraws from the oogonial wall and rounds off to form a single ovum. It has 8-nuclei. One of these migrates to the centre. The other seven nuclei become arranged along the periphery. There they gradually degenerate (E). The central one that survives enlarges to become the functional haploid female nucleus (F).

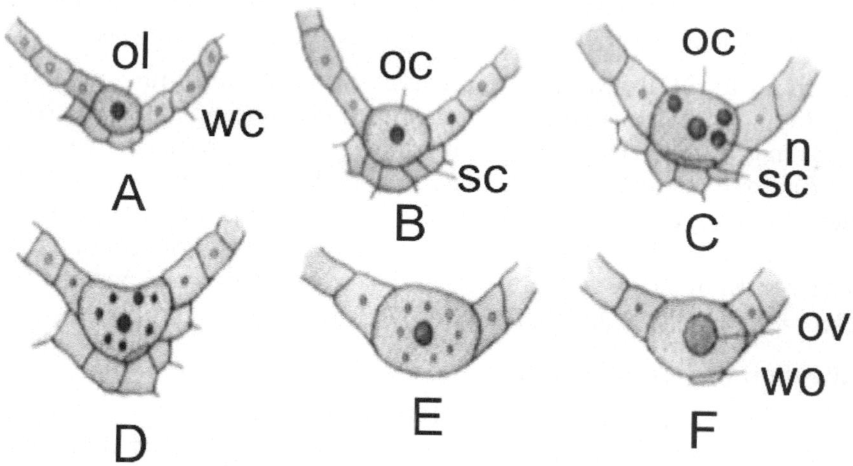

Figure 2.13: (A-F). Development of Oogonium in *Sargassum sp.* (A), Very early stage; *oi,* oogonial initial; *we,* conceptacle wall; (B), and (C), Formation of stalk cell (*sc*) and oogonial cell (*oc*) and the division of the nucleus (*n*) into four daughter-nuclei; (D), 8-nucleate stage; (E), One large functional nucleus and 7 degenerating nuclei; (F), Oogonium

Antheridia

The structure and development of antheridia is nearly similar to that of *Fucus*. They occur in large numbers and are so crowded in a single conceptacle that observation is made difficult. They are borne terminally usually on the lower branches of the scantily branched paraphyses. The paraphyses arise from the wall of the male conceptacle. The mature antheridium is a small, ovoid structure. It contains 64 biflagellate male cells or sperms. When sperms reach maturity the antheridia separate from their respective stalk cells by the action of water currents. They emerge from the conceptacle through its ostiole and float on the surface of water. The wall fo the antheridium gelatinises and sperms are liberated. The liberated sperm is a pear-shaped structure with a pointed anterior end. It has two laterally inserted flagella of unequal length. The longer flagellum is directed backward and the shorter is directed forward. The greater part of the body of the sperms is occupied by the large nucleus. The flagella arise from the blepharoplasts. The reported eye spot is derived from the chromatophore which is vestigial.

Development

The first antheridium arises from the cell of the wall of the male conceptacle. It grows into a papilla-like outgrowth. A transverse wall appears dividing it into two cells. The lower cell functions as the wall cell of the conceptacle. The upper cell again divides into two. Of these the lower is called the stalk cell and the upper antheridial cell. The stalk cell grows into a branch pushing the first terminal antheridium to one side. The branch cell may again divide into the terminal antheridium and basal branch cell. Finally the stalk cell forms one or two-celled paraphysis.

Each antheridial initial contains a single prominent nucleus and a few chromatophores. As it increases in size the antheridial nucleus divides repeatedly to form 64 daughter nuclei. The first two divisions constitute meiosis. The subsequent ones are mitotic. At the 32-nucleate stage the cytoplasm of the antheridium becomes divided by finely granular lamellae. The latter become prominent at the 64-nucleate stage of the antheridium. Consequently the mature antheridium contains 64 male cells each with one chromatopore. Each male cell is a pear-shaped structure. It develops two flagella of unequal lengths. They are inserted laterally.

Fertilization (Figure 2.14, A-D)

The liberated sperms swim towards the female conceptacles. Several of them surround each oogonium which is outside the ostiole still attached to the wall of the female conceptacle by the gelatinous stalk (A). The sperms become attached to the oogonial wall by their anterior flagella. Their posterior flagella lash violently in water. One of the sperms penetrates the wall of the oogonium and enters the ovum (C). The male nucleus passes in and comes to lie by the side of the female nucleus. The two finally fuse to form the zygote nucleus (D). Fertilization is thus affected while the ova are still contained in the oogonia which in turn are attached to the wall of the conceptacle.

Germination of Zygote (Figure 2.15)

Germination of zygote starts while it is still enclosed in the oogonium and the latter is still attached to the conceptacle wall. Eventually the oogonial wall gelatinises. The zygote is liberated. The liberated zygote rests on some solid object. There it divides by a transverse wall into an upper and lower cell (A). The lower cell gives rise to the fibrous rhizoid like processes and the upper forms the erect diploid plant.

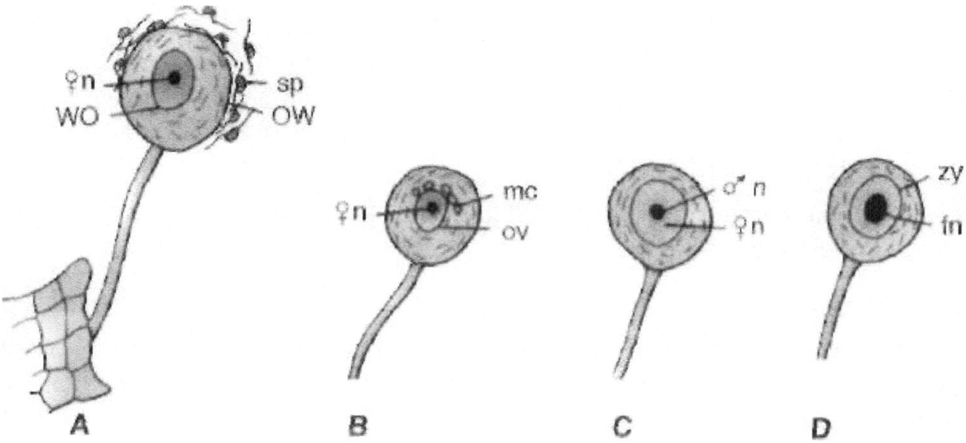

Figure 2.14: (A-D). Fertilization in *Sargassum sp.* (A), Sperm (*sp*) surrounding the oogonium; (B), Sperms having penetrated the oogonial wall; (C), Male and female nuclei lying side by side; (D), Act of fertilization accomplished; *fn,* fusion nuclei; *mc,* male cell; *sp,* sperm; *ov,* ovum; *zy,* zygote; *oiv,* oogonal wall; *fn,* female nucleus.

1. The conceptacle are always unisexual.
2. The oogonium is sessile and remains embedded in the wall of the conceptacle.
3. The oognoium contains only a single oosphere or ovum.
4. The oogonium after liberation remains attached to the conceptacle wall by a long gelatinous wall.

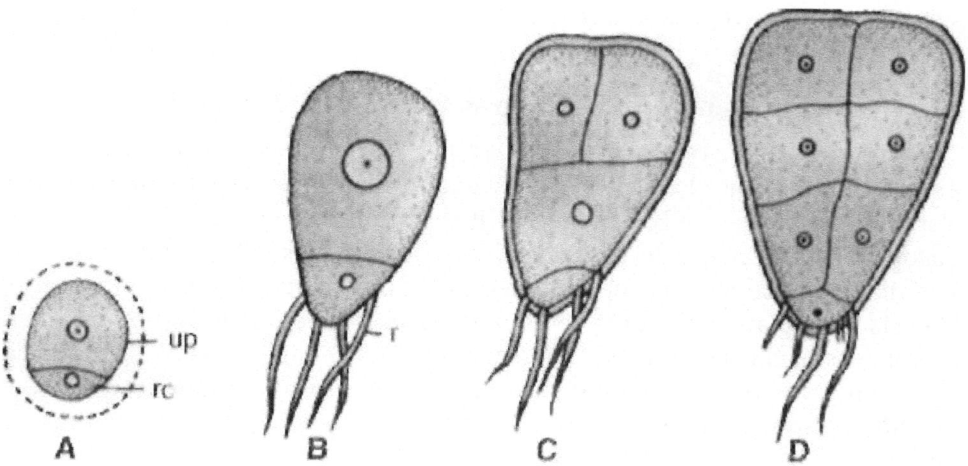

Figure 2.15 (A-D): Germination of Zygote in *Sargassum* sp. *up,* upper cell; *re,* rhizoid cell; r; rhizoid. Sexual reproduction in *Sargassum* differs from that of *Fucus* in the following respects:

5. The ovum is 8-nucleate at first, but 7 of these nuclei degenerate leaving behind only one nucleus at the time of fertilization.
6. The ovum or egg is fertilised while it is still contained in the oogonium which remains attached to the wall of the conceptacle. The fertilization is thus internal.
7. The zygote starts germinating while still attached to the parent by the gelatinous stalk.

Cryptoblasts

Besides the fertile conceptacle there are found in certain species *of Sargassum* sterile flask-shaped structures called the cryptoblasts (Figure 2.11, A). They are borne on the primary branches. They open to the outside by openings called cryptoblamata. The cryptoblasts do not contain any sex organs. They develop in the same way as the conceptacles and have the same structure. From the centre of the floor of the cryptoblast arises a group of colourless unbranched hairs. Each hair has a definite basal meristem (Figure 2.11, *bm*). These hairs project in tufts through the cryptoblamata. They fall off with age. Only the basal portions persist in the old cryptoblasts.

Intermingled with the colourless hairs are found coloured ones. In the fertile conceptacles sterile hairs are only present near the aperture. The cavity of the cryptoblasts is filled with mucilage secreted by the hairs. It is held that plants exposed to extreme conditions of the environment have the cryptoblasts developed in abundance.

Similar structure and development of the cryptoblasts and the conceptacles indicate that the two organs are homologous. This is confirmed by the fact that some people have reported the occurrence of abortive antheridia in cryptoblasts in certain cases.

Alternation of Generations (Figures 2.16 and 2.17)

Sargassum has a typical cyclosporian life cycle as in *Fucus.* There occurs only a single morphologically organised individual. It is the diploid sporophyte which is elaborate in structure and bushy in habit. It bears diploid sex organs (antheridia and oogonia) in flask-shaped depressions of the diploid thallus called the conceptacles. The thalli may be monoecious or dioecious but the conceptacles are unisexual. Meiosis takes place at the time of differentiation of the gametes from the diploid protoplasts of the sex organs. The gametes are thus haploid. The haploid male gamete fuses with the haploid female gamete to form the zygote in which the diploid condition is again resorted. The sexual or gametophyte generations in the life cycle has been completely suppressed. There is no morphologically organised gametophyte plant. The extremely reduced gametophyte is solely represented by the gametes only. From this we conclude that *Sargassum* exhibits no alternation of morphologically organised generations. There is, however, a cytological alternation

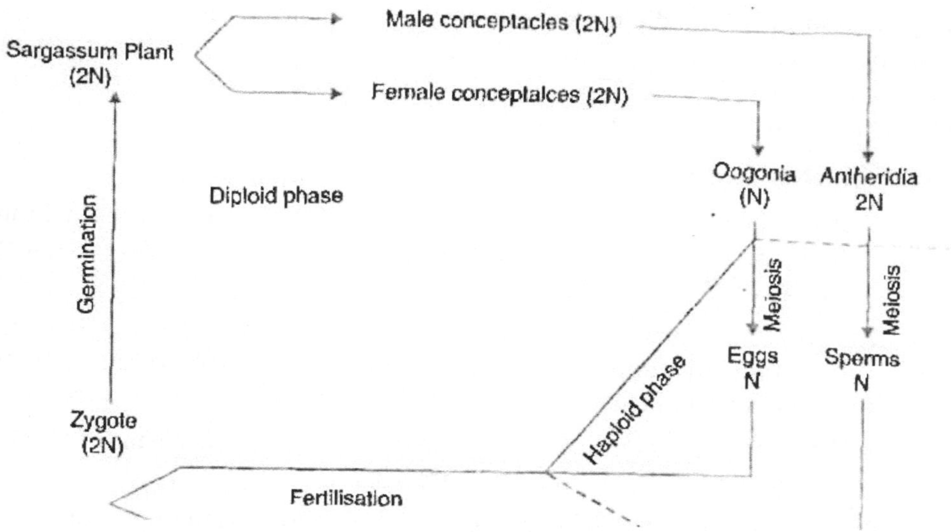

Figure 2.16: ***Sargassum*** **sp. Graphic representation of the life cycle.**

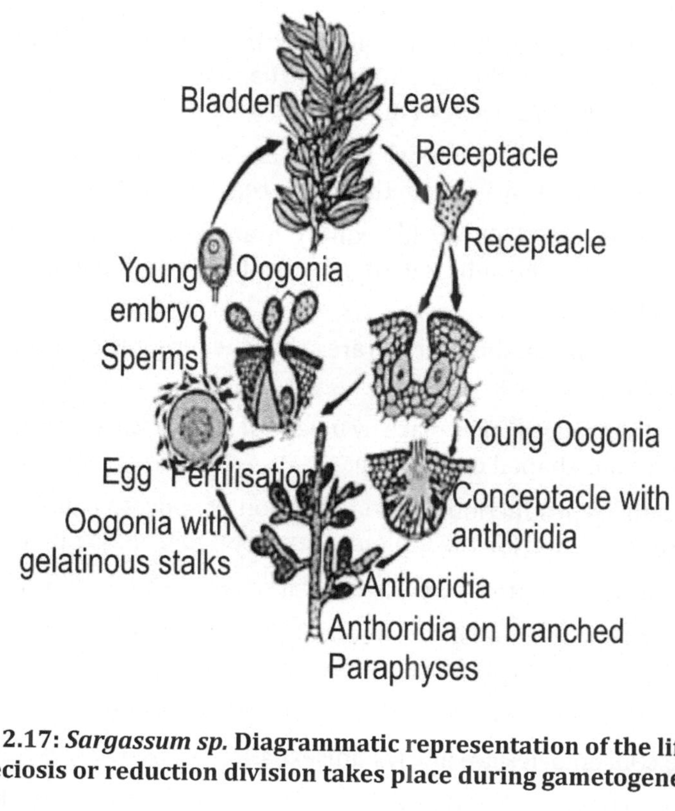

Figure 2.17: ***Sargassum sp.*** **Diagrammatic representation of the life-cycle. (Meciosis or reduction division takes place during gametogenesis).**

of generation from the diploid condition to the haploid at the time of gamete formation (meiosis) and the back to the diploid condition at the time of zygote formation (fertilization). The two critical points in the life cycle thus are meiosis and fertilization.

Points to Remember

1. The much branched, perennial diploid thallus apparently presents the appearance of a small, bushy angiosperm with its stem-like stipe, leaf blades and berry-like air bladders.
2. It is either cylindrical with the primary laterals arranged spirally and thus radially symmetrical or flattened with distichous arrangement of the laterals. In the latter case the thallus is bilaterally symmetrical.
3. The main axis is usually attached to the substratum by means of a well-developed holdfast with fibrous processes.
4. A few species occur in huge, free floating masses and multiply exclusively by the method of fragmentation.
5. The thallus is mainly composed of the much branched primary, long laterals.
6. The primary, long laterals generally bear the leafy shoot, secondary laterals of which the basal one is usually conspicuous and apparently subtends the branch system.
7. The primary and secondary laterals are sterile. The secondary laterals subtend branch systems in the upper region of the thallus.
8. The lower members of the axillary branch system may be reduced to air-bladders. The subsequent ones are fertile and may be cylindrical to flattened.
9. These cylindrical, fertile tracts are called the receptacles. They are studded with conceptacles.
10. Growth in length takes place by means of a three sided apical cell.situated in a funnel-shaped depression.
11. A sexual reproduction by the formation of spores is entirely absent—a feature unusual for the diploid, asexual or sporophyte plant.
12. It reproduces entirely by fragmentation and sexual reproduction.
13. Sexual reproduction is oogamous. The sex organs as in *Fucus,* are lodged in conceptacles which are confined to the receptacles.
14. Some species are monoecious and others dioecious.
15. The conceptacles are always unisexual.
16. The oogonia and antheridia are diploid structures—another unusual feature.

17. Meiosis takes place at the time of differentiation of the gametes.
18. The ripe oogonium has only one oosphere.
19. The oogonium is generally embedded in the wall of the conceptacle.
20. The antheridium produces 64 small, biflagellate sperms. The two flagella are of unequal length and inserted laterally.
21. Fertilization takes place when the oogonium is discharged but remains just outside the ostiole still attached to the wall of the conceptacle by means of a long, mucilaginous stalk.
22. The zygote germinates without undergoing any resting period and forms the diploid thallus.

SELF LEARNING MATERIALS

Phaeophyceae (Brown algae)

1. Introduction

- ☆ The brown algae are distinguished by their colour which varies from olive green through light golden to a rathr deep shade of brown.
- ☆ This is due to the presence of a golden brown Xanthophyll pigment Fucoxanthin in their Chromatophores.
- ☆ Brown algae reserve food in Mannitol and Laminarian are the photosynthetic product.
- ☆ The plant body is always immobile and multicellular. It is either a uniaxial branched filament or a large elaborately parenchymatous thallus orgnisation.
- ☆ The plant body is basel holdfast, a long or short stipe and an expanded blade.

2. Objective

On the basis of vegetative organisation and methods of sexual reproduction, division phaeophyta Kylin (1933) classified three classes on the basis of the type of life cycle:-

1. **Isogenerate**: It is characterized by the presence of two similar alternating generations.
2. **Heterogenerate**: There is an alternations of two dissimilar generations.
3. **Cyclosporae**: There is no alternation of generations.
   - ☆ The cells are uninucleate cell wall besides the cellulose and pectic substances contains a gel-like polysaccharide align and the carbohydrate fucoidin.

- ☆ Thallus is a heterotrichous filament consisting of well-defined erect and prostrate system.
- ☆ The growth in the rhizoidal portion is apical. The upright portion exhibits intercalary growth and in a few species it is trichothallic.

3. Content

- ☆ Majority of the forms are marine and only few are fresh water forms.
- ☆ Photosynthetic pigments in the Chromotophore and Chlorophyll a but no Chlorophyll b. β-Carotene and Xanthophyll such as Fucoxanthin and diatoxanthin. The Fucoxanthin mask the other pigmens.
- ☆ Presence of whitish granules termed as Fucosan vesicles in the cell in another characteristic of the brown algae.
- ☆ The reserve food is stored in the form of Mannital, Laminarian and rarely as fat droplets. They are formed in the pyrenoids, outside the chromatophores and diffuse into the Vacuoles.

4. Revision Points

- ☆ The fine, tufted, brownish tallus is a heterotrichous filament consisting of well defined erect and prostrate system.
- ☆ The main axis is usually attached to the substraction by means of a well development holdfast with fibrous process.
- ☆ In some species the mating gametes are similar (Isogamates) and sexual reproduction in isogamous other exhibit distinct anisogamy which may be physiological or morphological anisogamy.
- ☆ The zygote is not a resting body. It germinates immediately to give rise to the diploid plant.
- ☆ The haploid and diploid plants are morphologically alike. Ectocarpus thus exhibits an isomorphic type of alternation of generation.

5. Intext Question

1. Give an account of salient features of phaeophyceae and describe in brief the structure and reproduction in Ectocarpus.
2. Define alternation of Generations. Discuss the phenomenon with reference to the life cycle of Ectocarpus.
3. Enumerate distinguishing features of phaeophyceae. Describe male and female conceptacles of sargassum with suitable diagrams.
4. Give a diagrammatic account of the life cycle of Fucus. Illustrating the relative length of the haploid and diploid generations.
5. Write an account of Economic importanc of Brown algae.

6. Which of the following is called the Gulf weed?

   a) Sargassum b) Laminaria

   c) Fucus d) Padina

7. The alga commonly called rock weed is

   a) Ectocarpus b) Fucus

   c) Sargassum d) Laminaria

8. Sterile flask-shaped cavities present in the leaves of sargassum are called

   a) Conceptacles b) Cryptoplast

   c) Cryptoblamata d) Cryptostomata

9. The photosynthetic region in the thallus of sargassum

   a) Epidermis b) Cortex

   c) Medulla d) Cuticle

10. The product of photosynthesis in Sargassum is

    a) Starch b) Protein

    c) Manitol d) Glycerol

6. Summary

☆ The cells are uninucleate and the cell wall besides the cellulose and pectic subsances contain a gel-like polysaccharide align (alginic acid) and the Carbohydrate Fucoidin.

☆ The photosynthetic pigments in the chromatophore are chlorophyta but no Chlorophyll-b, β-Carotene and Xanthophyll such as Fucoxanthin and diatoxanthin. The Fucoxanthin masks the other pigments.

☆ The reserve food in stored in the form of Mannitol Laminarin. They are formed in the pyrenoids outside the Chromatophores and diffuse into the vacuoles.

☆ The dark brown or olive green thallus is diploid. It attains a considerable size and is differentiated into a disc like hold fast, stipe and frond.

☆ The stipe is usually a short stalk attached to the substraction by the holdfast and continues above into the midrib of the forked, flattened frond. The frond has either entire or serrate margin.

☆ The sex organs are lodged in flask-shaped cavities called the conceptacles which are developed at the ends of some of the branches of mature plants. These fertile, terminal tracts become swollen and are called the receptacles. These receptacles are studded with fertile conceptacles.

7. Supplementry Material

1) Journal of Phycology,
2) Phycologia,
3) Botanica Marina,
4) Phykos,
5) Indian Journal of Marine Science,
6) Botany for Degree Students "Algae"

8. Assignments

1. Give an accountof the life history of Fucus and discuss the question of alternation of generation in the genus.
2. Give an account of the habit and habitat of Fucus and explain how its form and structure is adapted for the environmental conditions.
3. Discuss the range of sexual reproduction in different genera of phaeophyceae studied by you.

9. Learning Activities

- ☆ Enumerate distinguishing features of phaeophyceae. Describe male and female conceptacles of Sargassum with suitable diagram.
- ☆ Describe sexual reproduction with the help of self-explanatory sketches, in Sargassum. Point out its differenes from Fucus.
- ☆ Give an account of salient features of Phaeophyceae and describe in brief the structure and reproduction in Ectocarpus.
- ☆ Bringout the role played by Sporangia and gametoangia in the life cycle of Ectocarpus.
- ☆ Give a brief illustrated account of structure and reproduction of any brown alga which you have studied.
- ☆ Write an account of Economic importance of Brown algae.

10. Key words

Unilocular Sporangia, Plurilocular Sporangia, Isomorphic alternation of generations, Heterotrichous, Diplohaplontic, Fucoxanthin, Alginic acid, Conceptacles, Receptacle, Gulf weed.

Rhodophyta

Rhodophyceae

General Characteristics Features of Rhodophyceae

1. Red algae is eukaryotic cell, pigments chlorophyll a, β-carotene and y phycoerythrin, chlorophyll d, alpha-carotene, lutein and y-phycocyanin.
2. Food reserve may occur as alcohols but are chiefly stored in the form of a polysaccharides as small granules free in the cytoplasm rather than in plastids and are known as florridean starch.
3. The occurrence of non-aggregated photosynthetic lamella or thylakoids with the plastids.
4. Growth of the Thallus may be apical as well as intercalary one of the stricking features of red algae, is the conspicuous cytoplasmic connection from cell to cell
5. Outstanding characteristic of this taxon is a total absence of any flagellated reproductive cell
6. Sexual reproduction is of most advanced oogamous type.
7. Male sex organs are known as spermatangia and female as carpogonia. carpogonium is a flask shaped structure having a long neck like trichogyne.

Classification

Rhodophyta

Bangiophycideae	Florideophycideae
(Absence of Pit connection)	(Presence of Pit connection)
ORDER	
1. Porphyridales	5. Nenolionales
2. Goniotrichales	6. Cryptonemiales
3. Compsopogonales	7. Gigartinales
4. Bangiales	8. Rhodymeniales
	9. Ceramiales

Batrachospermum

Systematic Position

Division : Rhodophyta (Rhodophycophyta)

Class : Rhodophyceae

Sub Class : Florideae

Order : Nemalionales

Family : Batrachospermaceae

Genus : *Batrachospermum*

Characteristic Features

1. It is one of the few, inland fresh water red algae commonly found in tropical, sub tropical and temperate regions.
2. It occurs in slow flowing streams, water falls, margins of lakes, in springs usually under the shade attached to rocks.
3. The thallus is bluish or olive green or violet green in colour when the plants are growing in shallow water and dark violet or reddish colour in deep water specimens.
4. The adult thalli are uniaxial in its simplest form, heterotrichous, may attain a length of 20 cm of even more.
5. The branches (or) laterals it bears are of two kinds
    i. Primary laterals or branches of limited growth
    ii. Long branches or branches of unlimited growth
6. The growth of axial filament and branches takes place by means of apical cell.
7. The sexual reproduction is of a advanced oogamous type by formation of male sex organs antheridia or spermatangia and female carpogonia.

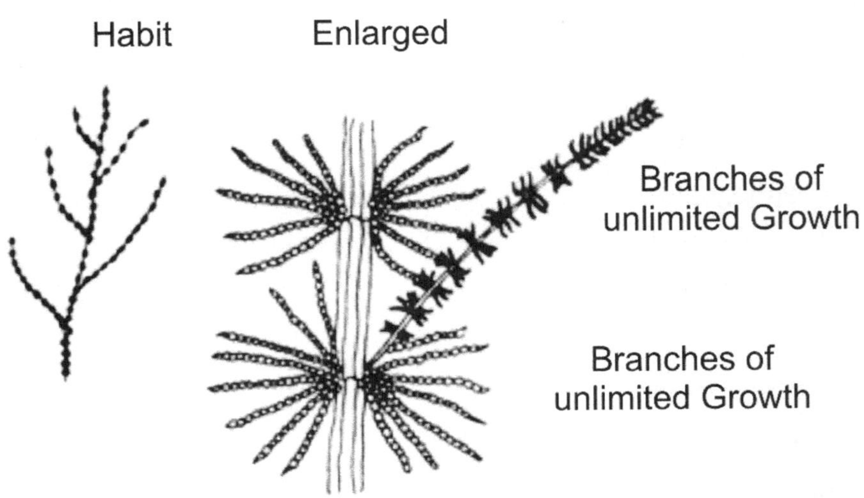

Figure 2.18

8. The zygote developed as a result of fertilization undergoes meiosis to form haploid gonimoblast filament, which in turn gives rise to cystocarp or carposporophyte on parasite on parent.
9. The cystocarp produces a haploid carpospore, germinate to form protonema like heterotrichous structure called chontransia stage and later develops into adult *Batrachospermum*.
10. Asexual reproduction is brought about by means of non-motile, uninucleate spores called to monospores.
11. Cytologically the life cycle of *Batrachospermum* is called haplobiontic and morphologically as triphasic of trigenic or haplohaplohaplontic

Polysiphonia: Greville, 1824

(Poly – Many, Siphonia – Siphons)

Systematic Position

Division	: Rhodophyta (Rhodophycophyta)
Class	: Rhodophyceae
Sub Class	: Florideae
Order	: Ceramiales
Family	: Rhodomelaceae
Sub family	: Polysiphonieae
Genus	: *Polysiphonia*

Characteristic Features

1. The plant body is polysiphonous in nature consisting of the document and the upright portion.
2. The feathery, upright portion consisting of a much branched system of relatively large branches bearing numerous small branches called the trichoblast. The later bears the sex organs.
3. Occurrence of pit connections between the adjoining central cells and between the central cells and the pericentral cells, the characteristic feature of red algae.
4. Occurrence of three kinds of plants-gametophytes, carposporophyte and tetrasporophyte.
5. Sexual reproduction is oogamous. The male sex organ are called the spermatangia and the female carpogonia (flask shaped)
6. The diploid nucleus formed by fertilization lies at the basal swollen portion of the carpogonium. The latter communicates by means of a tubular connection with the auxillary cell cut off from the supporting cell.

7. The development of gonimoblast filaments forms the auxillary cell.
8. The formation of a diploid carpospore in the modified terminal cell of the gonimoblast filament now called the carposporangium bearing the carpospores.
9. Development of the tetrasporophyte by the germination of the diploid carpospore, is concerned with production of meiospores called tetraspores (haploid)
10. Germination of haploid tetraspores to gives rise to the gametophyte or sexual plants concerned with sexual reproduction.
11. The life cycle with three phases in which one is haploid and two diploid is called diplodiplohaplontic or diplobiontic.

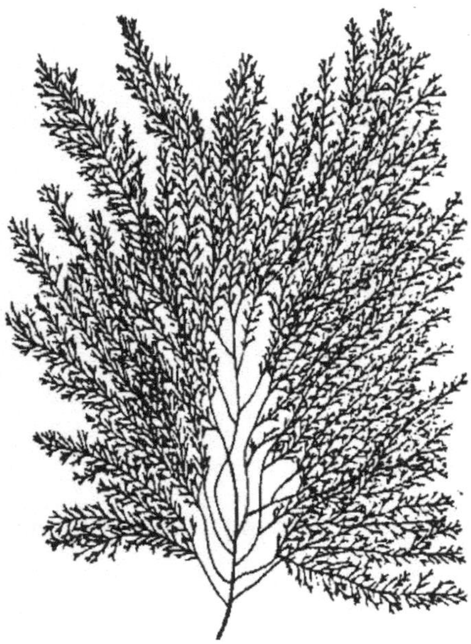

a) Habit

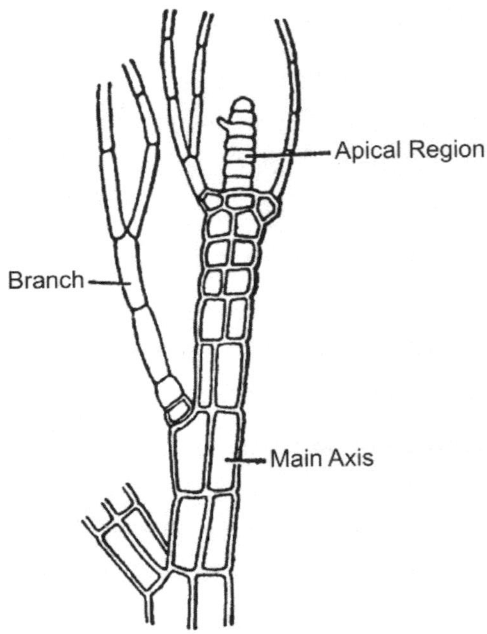

b) Portion Enlarged

Figure 2.19

GRACILARIA

Systematic Position

Division : Rhodophyta
Class : Rhodophyceae
Order : Gracilariales
Family : Gracilariaceae
Genus : *Gracilaria*

Characteristic Features

1. Plants attached to the substratum by the thin spreading disc.
2. Fronds 5-10 cm high, 2-4 mm wide, compressed, except near the base where it is subserete.
3. The branches given off from the edges of the flat thallus either alternately or 2-3 seriately from the same side.
4. The upper parts thus getting an antler like appearance upto 4 mm broad below the ramification, apics acute or Bifid.
5. Medullary tissue of large cells upto 250 mm diameter
6. Cystocarps hermispherical, prominent and profusely scattered on the flat side of the thallus.
7. Spores rounded, cartilaginous consistency
8. The transverse section of the thallus shows the outer 1-2 layers of epidermis middle 2-3 layers cortex and inner 1-2 layered medulla.

a) Habit

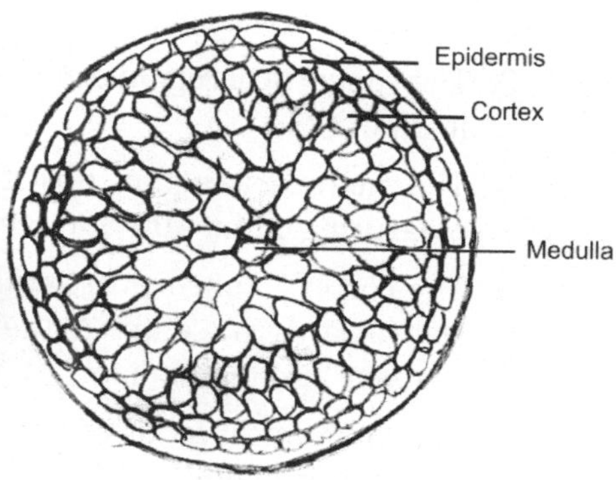

b) T.S. of Thallus

Figure 2.20

Hypnea

Systematic Position

Diyision	: Rhodophyta
Class	: Rhodophyceae
Order	: Gigartinales
Family	: Hypneaceae
Genus	: *Hypnea*

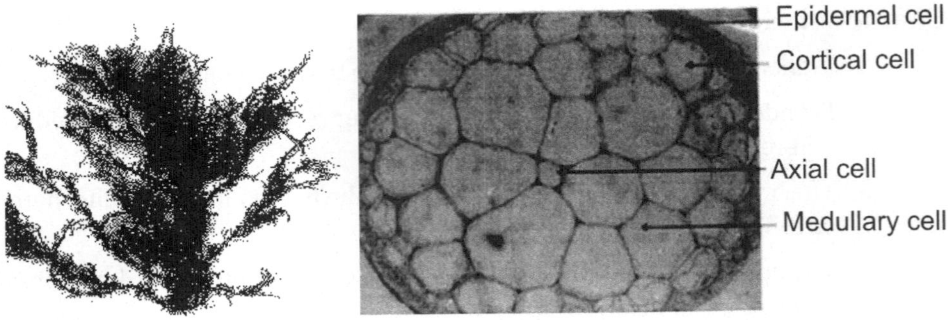

a) H. *musclformis* - Vegetative Plants **b) Cross Section of *Hypnea***

Figure 2.21

Characteristic Features

1. Thallus shows a uniaxial construction.
2. The outer cortical region consists of a single layer of small densely pigmented cortical cells. Each cell is about 10-25 mm diam.
3. The subcortical region consists of 1-2 layers of cells. Cells are 30-50 mm in diam.
4. The medullary region consists of 2-3 layers of larger cells. Each cell is measuring about 100-300 mm diam.
5. The cortical, subcortical and medullary cells are progressively large in size towards the centre. At the centre a distinct axial cell is present which is about 25-40 mm diam.
6. The anatomical features of the vegetative portions of cystocarpic and tetrasporic plants are similar to that of the vegetative plants.

I. Uniaxial Form with Procarps

Gigartinales: *Rhodophyltidaceae*

Cystoclonium purpureum

The terrete, somewhat cartilaginous thallus, which is often dull purple, in. **colour** and may reach a length of 60 cm. has a prominent main axis bearing numerous richly branched laterals Certain branches may end in tendrils. A minujte **disc** primarily anchors the plants, but the base of the axis later gives rise to a number of stolons (st), which become attached by squat haptera (h) and produce further erect shoots. The individuals commonly last for one season only although some perennate by means of the stolons.

The anatomical structure of the Cystoclonium purpureum showed the following special features. The segments of the two-sided apical cell cut off two pericentrals on their higher side, which is situated, alternately to right and left. The two pericentrals lie in perpendicular planes. The rest of the segment constitutes the axial cell. Such asymmetrical development of pericentrals takes place in many of the more specialised Florideae. The axial cells undergo marked elongation, while the pericentrals as usual give rise to richly branched laterals, which form the compact cortical region (Figure 2.22).

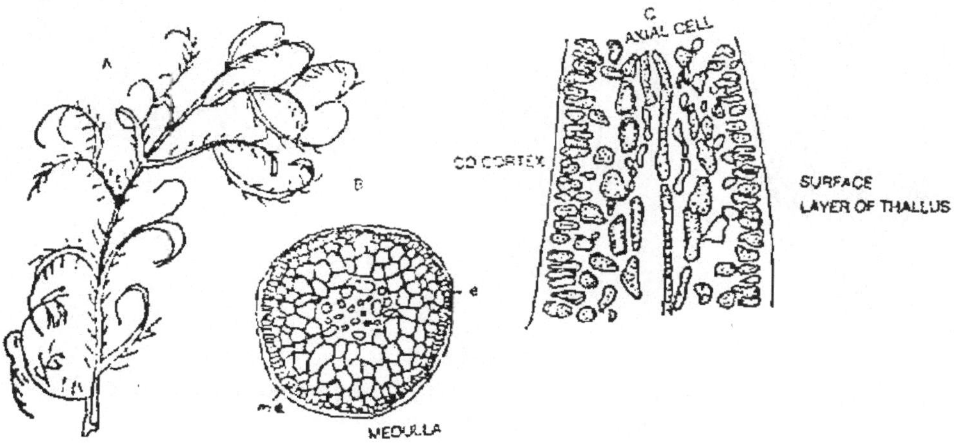

Figure 2.22: *Cytoclonium purpuram*
A) Habit, B) Transverse section of mature branch, C) Longitudinal section of the same. With Procarps: Gigartinales: Gracilariaceae: *Gracilaria* and its species: (a) *G. confervoides*; (d) *G. lichenoides* and (c) *G. compressa*.

A different structure is found in *Gracilaria*: of which *G. confervoides* is a **widely** distributed northern lithophyte found at low-tide level and below, although **a** characteristic inhabitant of sandy bottom in the Mediterranean. Its dark red terete fronds, which are up to 50 cm long, bear more or less numerous tapering branches and several arise from a large fleshy disc composed of closely compacted vertical threads; this, as well as the greater part of the erect thallus, persists from one season to the next. Other species have more or less compressed thalli. The mature thallus shows little anatomical differentiation, *since all* the internal cells (medulla) are *large and* isodiametric and an axial thread is unrecognisable. The small cells forming the narrow cortical zone contain a number of ribbon-shaped chromatophores. Unicellular hairs arise from enlarged peripheral cells, which possess abundant conrents and later become multinucleate (Figure 2.23).

The instances hitherto considered in this section show how, despite an -essentially similar apical structure, the primary uniaxial construction might become obscured among Gigartinales, either by the development of accessory longitudinal threads or by equal enlargement of all the internal cells (*Gradfaric*). Other uniaxial Gigartinales show marked flattening of the thallus.

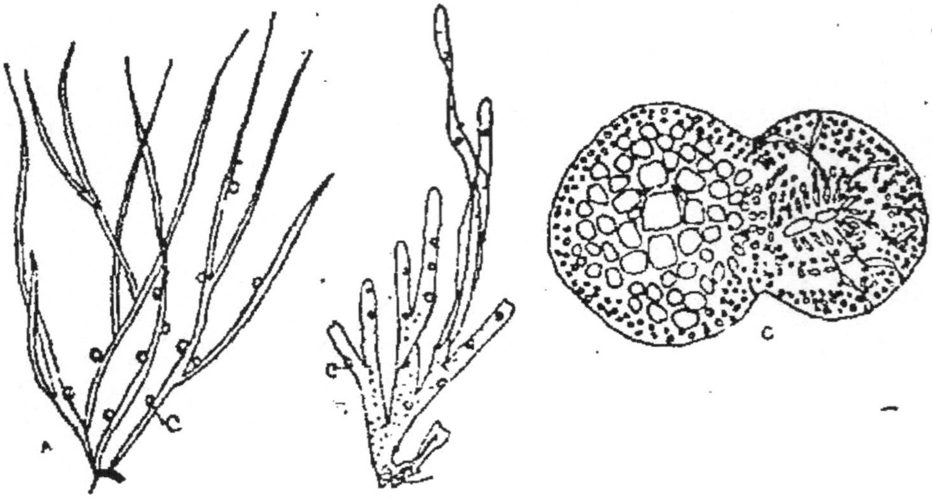

Figure 2.23
a) *G. confervoides*, b) *G. compressa*, c) *G. confervoides* transverse section of thallus with cystocarp.

II. MULTIAXIAL FORM

(a) *Ahnfeltia plicata*

(b) *Chondrus crispus*

Ahnfeltia plicata, a widespread, slow-growing perennial lithophyte recognised by the wiry texture of the dark purple thallus, is attached by a broad disc composed of vertical rows of almost cubical cells with firm walls, Numerous closely aggregated threads occupy the apex. In the mature thallus the extensive medulla consists of elongate thick-walled cells, no doubt fulfilling a mechanical function and merging into a cortex composed of small, radially arranged, elements. The older parts, especially in plants exposed to rough seas exhibit extensive secondary thickening which takes place periodically by division of the peripheral cells of the cortex and results in the formation of a number of successive strata, consisting of cubical, radially arranged cells (Figure 2.24).

Similarly, the morphological and anatomical structures of *Chondrus* afford **clear** evidence of a uniaxial construction (Figure 2.25).

The Special Development of the Multiaxial Type among Rhodymeniales

A multiaxial construction of a distinctive type is found in the genera *Champia, Chytcdadia,* and *Lomentaria,* which belong to the Champiaceae of Rhodymeniales.

Champia paruula a widely distributed epiphyte in the warmer seas of the Northern Hemisphere forms dense bushy tufts arising from a number of creeping

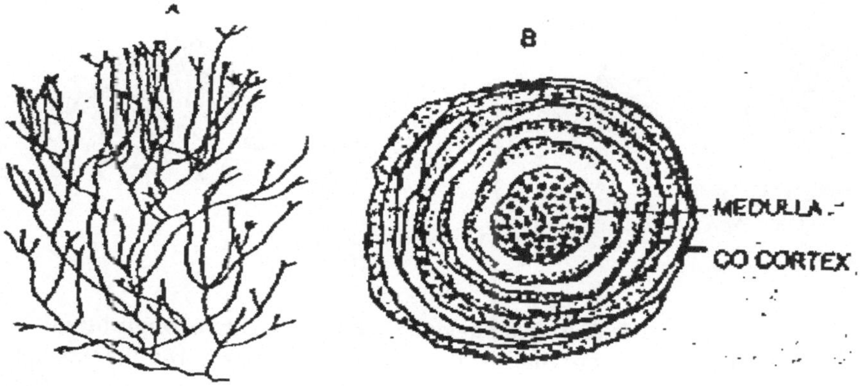

Figure 2.24
a) Habit, b) Transverse section of an older branch; showing secondary thickening.

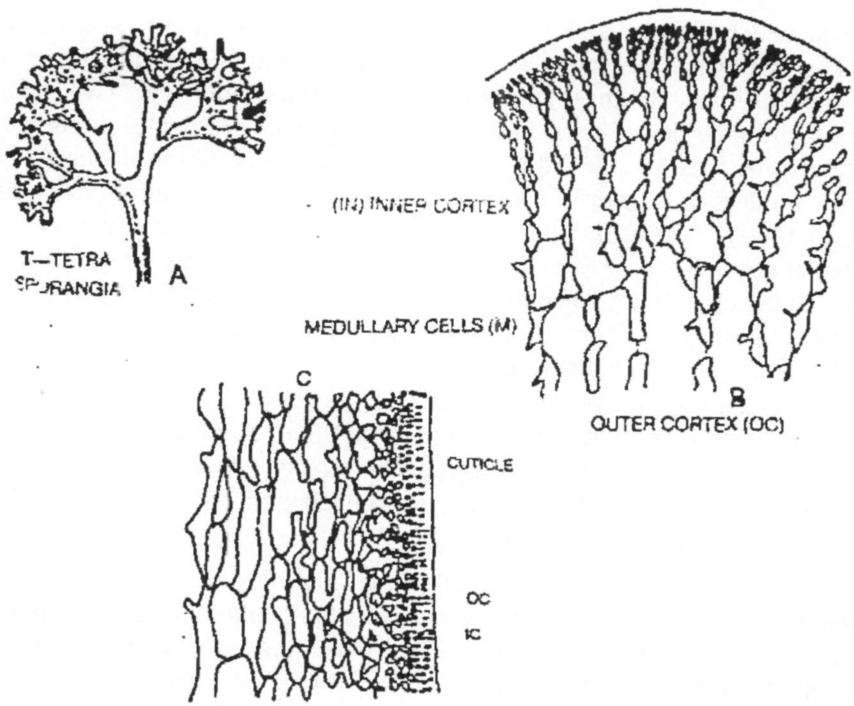

Figure 2.25: A-C: *Chondrus crispus;*
A, Habit; B, Transverse and C, Longitudinal section of mature thallus.

threads attached by multicellular haptera; the branches here often arise singly. The more robust and little-branched C lumbricallis has a wall of a number of layers of cells, while the diaphragms may be several layered (Figure 2.26).

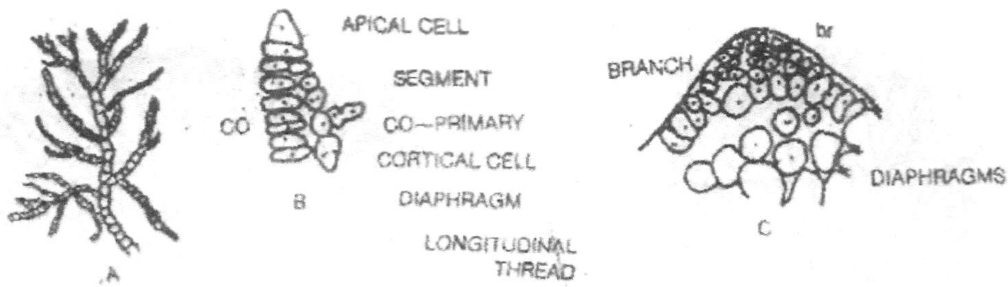

Figure 2.26: A-C: *Champia parvula;* A, Habit; B, Single longitudinal section through apex of thallus; C, Transverse section of thallus through the point of origin of a a branch (Br). a, apical calls; Co primary cortical cells; cu, cuticle; d, diaphragms; l, bulb-cells; l, longitudinal threads; m, mucilage; r, rhizoids; s, sc, secondary cortical cells.

The fistular thalli, which are attached by a minute disc, are constricted at the points at which the whorled branches arise; both here and at the bases of the. branches; diaphragms bridge the central cavity, which is elsewhere filled with mucilage. The entire thallus is encased in a mucilage envelope with a well-marked cuticle. In the older parts, the surface consists in 'the main of a single layer of radially elongated cells, with 16 to 20 little - branched threads (1) running of longitudinally *on* their inner side at more or less even intervals. These longitudinal threads converge towards the rounded apex, where they terminate in apical cells, not all situated at quite the same level. Each apical cell cuts off a single row of segments, which divide tangentially. The outer halves constitute primary cortical cells, while the inner gradually lengthen to form a longitudinal thread. The primcoy cortical cells divide anticlinaly to form the peripheral layer of large cells, which contain several nuclei, numerous very small chromatophores, and in some localities light reflecting bodies. Later they cut off obliquely on their outer side a certain number of small cells forming a disconnected superficial layer, which is stated to be most extensive in plants exposed to strong light. Especially in youngest individuals its cells produce elongate, thick-walled hairs.

The cells of the longitudinal threads undergo no further transverse division, but most of them cut off on their inner side a small spherical cell with plentiful contents, the bulb-cells of American writers. At the points of branching of the thalli these cells divide to form threads, which unite, with the establishment of secondary pit-connections, to produce the one-layered diaphragms. The numerous bulb-cells that do not develop into diaphragms have been interpreted as glandular elements or rather more usually, as undeveloper diaphragms. Branches originate by the outgrowth of several peripheral cells into threads, which unite to form the apex of the new axis; adventitious branches appear at indeterminate places in older plants.

The Special Developments of the Uniaxial type among Ceramiales

The numerous genera of Ceramiales exhibit great diversity in vegetative construction, as exemplified by the families Ceramiaceae, Delesseriaceae, Rhodomelaceae and Dasyaceae, although these only present minor differences in their reproduction. Diverse genera attain to a high degree of morphological develop a single bipolar axis from the first and all traces of heterotrichy are lacking, while the first septum is usually described as running parallel to the substratum.

a. Ceramiaceae: *Ceramium deslongchampil*

b. Delesserieae: *Delesseria sanguina*

c. Rhodomelaceae:

   (i) Rhodomelaceae with radial construction Polysiphonieae: *Polysiphonia* sp.

   (ii) Rhodomelaceae with bilateral and dorsirvental construction. Lopbosiphonieae: *Stictothamnion* sp.

Reproduction

When gametes fuse to form a zygote the chromosome number is doubled and before gametes can again be produced the original haploid number must be restored. In some algae this occurs when the zygote germinates; in other algae a diploid phase is introduced and from such a phase a haploid spore is produced. In certain algae either phase is capable of reproducing itself and not giving rise to the alternate generation; hence the use of the term life history rather than life cycle. In those algae where there is only a diploid generation (Fucales, Siphonales) reduction division occurs during gamete formation and fusion of the gametes restores the double number of chromosomes.

Development of Lateral of Ultimated Growth on Trichoblast (t) Branches and Pericentrals

Svedelius, in his studies on the Rhodophyceae, coined a number of terms, which have subsequently come into common use for the algae:

1. Haplont: A sexual haploid plant with only the zygote diploid (Volvox -Chlorophyceae).
2. Diplont: A sexual diploid plant in which only the gametes are haploid (Codium - Chlorophyceae)
3. Haplobiont: A plant possessing cytologically only one kind of individual in the life cycle; this may be either a haplont or a diplont.
4. Diplobiont: A plant possessing cytologically two kinds of individuals in the life cycle, and hence an alternation of generations.

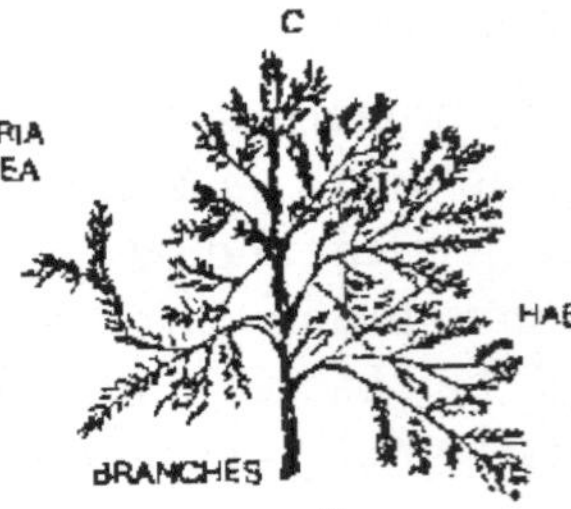

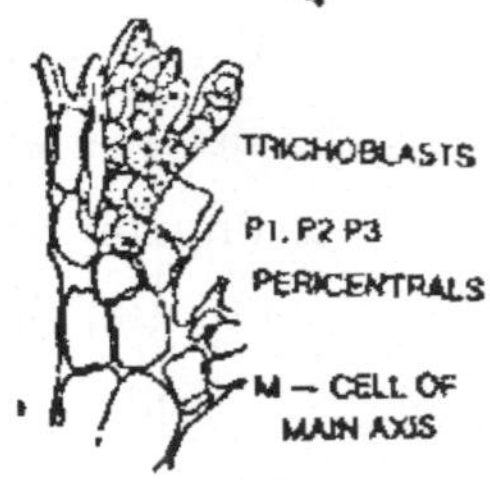

Ceramium sp. B, *Delesseria sanguinea*. C, *Polysiphonia*.

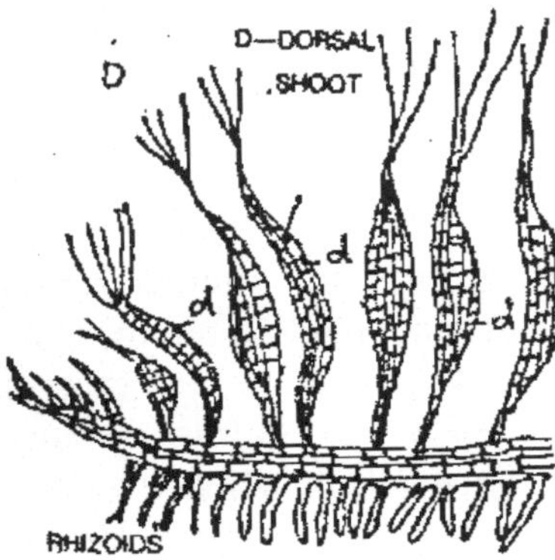

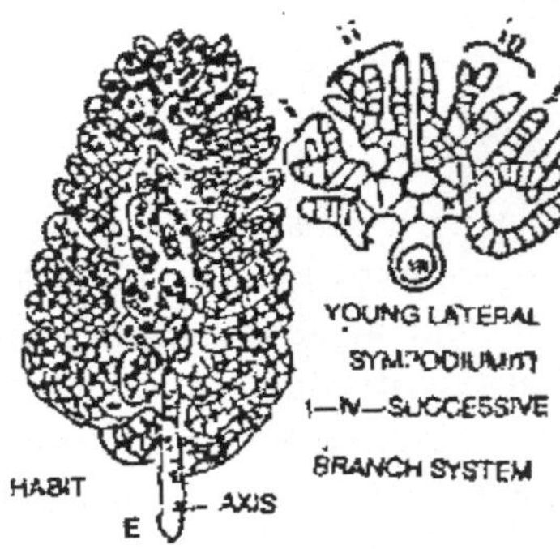

D, *Stictothamnion* sp.; E, *Dictyurus* sp.

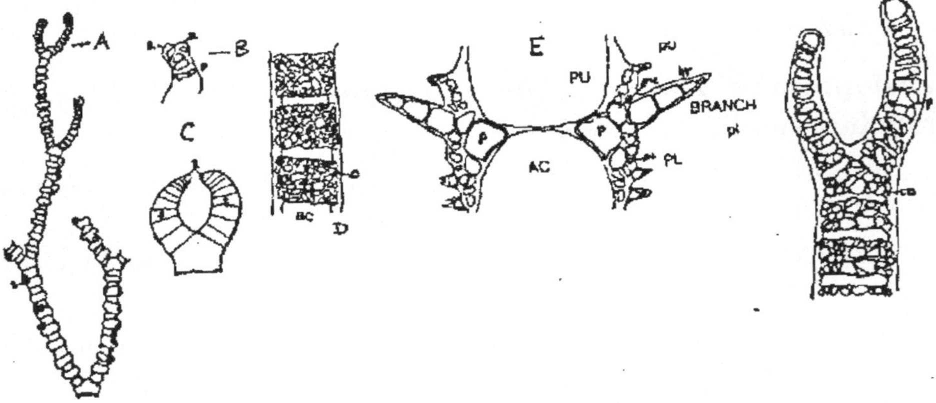

Figure 2.27: A-D; Ceramium sp, A apex of a banch-system; B, apical branching, C, diagram of apex; D, structure of a mature thallus branch. E, mature thallus branch with dichotomy; F, Development of cortication and branches, describes and exhibits the special development of the uniaxial type among Ceramials.

Referring Coleochaete, (Chlorophyceae), where the oospore (zygote)- in germination divides into 8 or 16 cells, each of which produces a swarmer. A similar condition is met among Rhodophyceae in the Bangiales and Nemalionales (Haplobiontic forms), in the latter group the fertilized carpogonium puts forth a number of short filaments (gonimoblasts), some of whose cells produce motionless spores (carpospores) which give rise to a new thallus. In all of these cases, however, despite the more extensive development of the cell resulting from sexual fusion before it gives rise to the spores producing the haploid plant, the first division that takes place in the zygote brings about reduction of chromosome number.

In the regular alternation of two outwardly identical individuals, the one is haploid bearing gametes; the other is a diploid producing asexual spore. Reduction occurs at the end of the diploid phase during the formation of spores and the zygote grows directly into the asexual individual. It is further characteristic of all the Rhodophyceae, apart from the Bangiales and Nemalionales, but in this case there are two diploid phases alternating with the haploid one (diplobiontic forms). The asexual diploid individuals bear tetraspores, during the formation of which reduction occurs and these spores give rise to a perfectly similar haploid individual bearing the sex organs; the zygote, however in this case does not immediately give rise to- the ordinary asexual individual but with diverse complications produces gonimoblasts representing an additional diploid phase, from which carpospores are formed. It is these carpospores that grow into the diploid individual bearing the tetraspores.

The origin of forms with homologous alternation from the haploid type must have occurred by a sudden mutation, the zygote retaining the diploid condition and, without the immediate spore production found in the haploid types, giving rise direct to a new individual which being diploid could only reproduce asexually.

Sexual reproduction in Rhodophyceae is invariably oogmous. The female organ (carpogonium), which usually bears a special receptive process (trichogyne), exhibits retention of the single ovum (egg) and the latter shows no contraction either before or after fertilisation. The male organ (antheridium) produces a single motionless male cell (spermatium), which is conveyed passively to the trichogyne. The zygote, either by direct division as in Bangiales or after the production of filamentous outgrowths (gonimoblasts) as in Nemalional gives rise to a number of sporangia, each forming a single, naked spore (carpospore). In most Florideae, however, the gonimoblasts do not arise directly from the zygote^ but from other cells (auxiliary cells), which belong to the same branch system as the carpogonium or are more or less widely, separated from it. Intimate combinations of carpogonia and auxiliary cells are termed procarps, and in them there is direct fusion between the carpogonium and the auxiliary cells, while when the latter are spatially separated, the products of the zygote nucleus are conveyed *to* them by special connecting filaments, frequently elongate and usually several in number. A descendant by division of the zygote nucleus, gonimoblasts producing carpospores arise from the auxiliary cell. In many Florideae there is a marked fusion of cells during the development of the gonimoblast, a process connected with the supply of nutrient to the maturing carpospores.

Reduction occurs either at the first division of the zygote nucleus (Bangiales, Nemalionales) or it is postponed and take place in special tetrasporangia, borne on individuals distinct from, but morphologically identical with, the sexual plants (other florideae). In Bangiales and Nemalionales there are thus two haploid phases the sexual individual and the life cycle is diplobiontic. With few exceptions the tetrasporangia form four spores, which exhibit three types of arrangement within the sporangium (cruciate, zonate, tetrahedral). Direct reproduction of the sexual phase by means of accessory spores (usually monospores) is met with especially in Bangiales and Nemalionales.

Many Florideae possess distinct male and female plants, although several Nemalionales and Corallinaceae, *for* example are monoecious. Not uncommonly the male plants are smaller than the female and they often perish altogether or almost entirely after liberation of the male cells. Their frequent ephemeral nature probably accounts for their rare discovery in diverse Florideae. In *Griffithsia* Lewis found groups of four plants with interwoven rhizoids, presumably the products of germination of the four spores of a tetra sporangium; two of these plants were male and two female, implying genotypic sex determination.

(a) The Female Organs

The female organ or carpogonium consists of a dilated basal portion harbouring the female cell and a usually narrow elongate process (t), the trichogyne, which constitutes the receptive organ for the male cells and commonly has a very gelatinous membrane. The female protoplast, which contains some reserves, exhibits no contraction either before or after fertilisation. In most Florideae the carpogonium terminates a short, often 3-4 celled, lateral, spoken of as the carpogonial branch; the cell (su) from which this branch arises is termed the supporting cell. In forms with a compact thallus the carpognial branch usually arises from an internal cell, and it is the function of.the trichogyne (t) to maintain connection with the external medium. Such carpogonial branches, however, often ‘ arise near the growing tips from cells situated close to the surface although as growth.progresses the supporting cell becomes deeply embedded. In the less specialised forms the lower cells of the carpogonial branch often bear laterals but these are lacking in Rhodymeniales, Ceramiaies, and most Gigartinales. In some members of these orders the cells of the carpogonial branch contain several nuclei and this may indicate arrested branch formation. In diverse Nemalionales the carpogonial branches are part of the normal branch - system, but in most Red Algae they constitute accessory laterals. The cells of the carpogonial branches are commonly colourless (Figure 2.28).

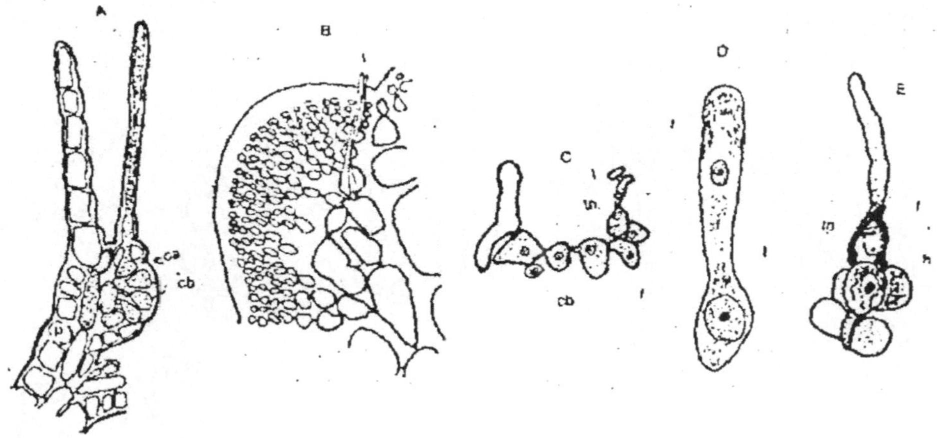

Figure 2.28: Carpogonia of Florideae: A *Rhodomela virgata* Plocamium coceineum; *C. Dumontia increassata: Polysiphonia,* E. *Scinaia furcellata* C, axial cell; cb, carpogonial branch; cp carpogonium N, nucleus of female cell; h, hypogynous cell; p,c pericetra;s; sc sterile cell; su, supporting cells t, trichogyne tn, trichogyne nucleus.

(b) The Male Organs

The antheridia of Florideae are spherical or oblong, unicellular structures, producing a single male cell or spermatium from the entire contents. As a rule they arise in large numbers in close proximity to one another so as to form white or pale coloured dots or patches, visible to the naked eye. Several antheridia often 3, sometimes 2, 4, or 5 are produced successively from a common mother cell, which is often elongate. The mother cells, which usually terminate short, often densely compacted, branch - systems and are invariably uninucleate, commonly possess reduced chromatophores and are sometimes altogether colourless. They often form a palisade like layer, especially in the more compact types (Figure 2.29)

As a general rule the antheridia first appear as protuberances which usually arise sub terminally and successively from different sides of the mother cell. As an antheridia protuberance develops, the nucleus shifts to the point of orgin and enters into early prophase. After completion of mitosis one daughter nucleus enters the antheridium, while the other again passes to the base of the mother cell, and this function may repeat again thus producing more antheridium. The mature antheridium commonly possesses a thick wall the spermatium is liberated through a narrow apical slit it in the elastic wall.

(c) Fertilization

The spermatia probably often have approximately the same specific gravity as the surrounding medium so that they remain floating for some considerable time.

Their large numbers ensure contact with a trichogyne, the tip of which is evidently more or less mucilaginous.

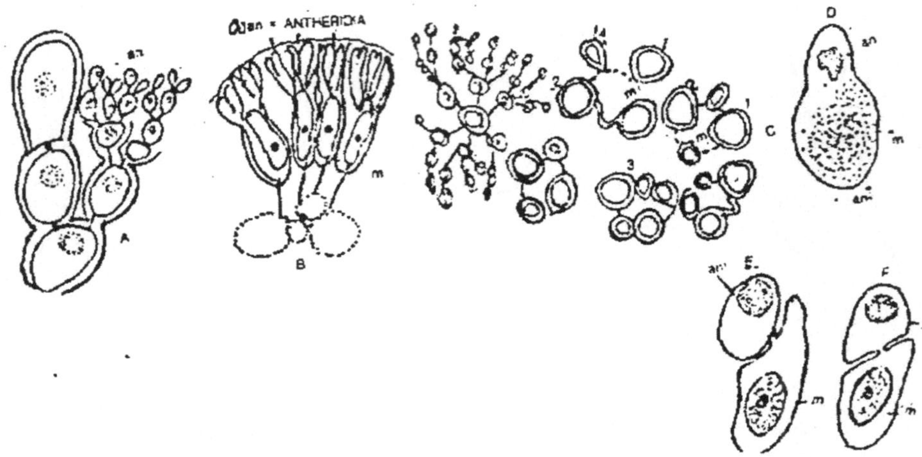

Figure 2.29: Antheridia and antheridial development in Florideae. A, *Helminthocladia;* B - F *Polysiphonia violacea;* B, Antheridia in cluster; C, seen from the surface; D - F, stages in formation of antheridia.

In Rhodophyceae the spermatia perform slow gliding movements similar to those reported for the carpospores and tetraspores or certain Florideae, and these are perhaps of wider occurrence. They will facilitate contact with trichogynes that do not project markedly or are embedded in mucilage. At the point of contact both the membrane of the spermatium and the wall of the trichogyne are dissolved, leaving an open passage for the nucleus of the male gamete.

In *Batrachospermum* and *Nemalion* the spermatium nucleus ordinarily divides after contact with the trichogyne, although only one nucleus fuses with the female. While first described in *Nemalion* the passage of the male nucleus to the basal part of the carpogonium, the apposition of the two sex nuclei, and their fusion. The last process has not been observed, but indirect evidence of the production of a diploid fusion nucleus is considerable, and in a large proportion of Florideae fertilization no doubt follows on contact of a spermatium with a trichogyne. As the male nucleus advances, the cytoplasm of the trichogyne not uncommonly shows signs of disorganisation, exhibiting numerous particles staining deeply with haematoxylin, whilst the part in front is still homogeneous the female nucleus seems frequently to move a short distance upwards to meet the male (Figure 2.30)

After penetration of the male nucleus the trichogyne becomes separated at its base from the rest of the carpogonium this is commonly affected by progressive thickening of the membrane until complete closure occurs, although sometimes the channel seems to be blocked by a plug of membrane substance. The trichogyne either persists for some time or rapidly withers.

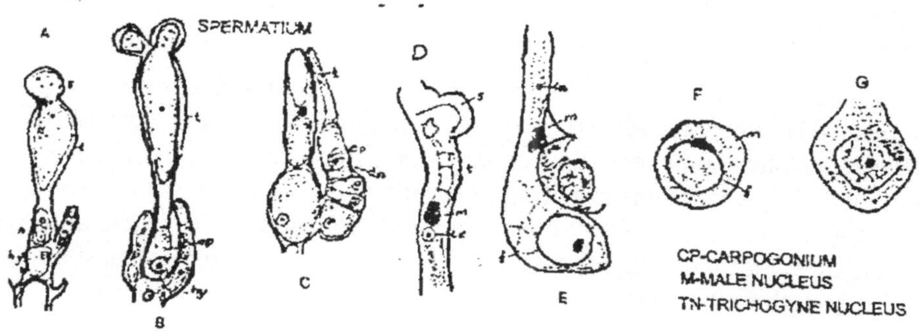

Figure 2.30: Fertilisation in Florideae: A, B, Batrachospermum, *moniliforme,* post fertilisation stages: C, *Helminthora divaricata.* Post fertilisation, D - G *Polysiphonia* Violacea, successive stages in passage of male nucleus and nuclear fusion.

d) Post-fertilisation Changes and the General Features of the Life Cycle

In Nemalionales the first nuclear divisions within the fertilised carpogonium are meiotic but in most Florideae they are mitotic, reduction being postponed to the time of tetraspore formation. The fertilised carpogonium invariably puts forth one or more threads, the ultimate function of which is the production of asexual spores (carpospores), but the manner of formation of the latter varies in the different orders. In Nemalionales the haploid threads or gonimoblasts arising from the carpogonium, produce the carpospores directly. The threads are relatively short and well branched so as to form a dense cluster, and certain enlarged cells, filled with cytoplasm and food-reserves, constitute carposporangia, from each of which a single naked carpospore is liberated. The carpospore gives rise to a plant which again bears sex organs so that in the life-cycle of Nemalionales the zygote constitutes the only *diploid* stage, while there are two haploid phases, the ordinary seaweed and the gonimoblast - threads originating from the zygote. Such Florideae have been designated haplobiontic.

In the remaining Florideae the threads, arising from the fertilised carpogonium, contain diploid nuclei and are formed singly or in small number. They grow towards and establish contact with certain cells, which are situated near the end or at a more or less considerable distance and are often specially differentiated by size or rich contents; in certain orders these auxiliary cells are clearly recognisable before the threads from the carpogonia reach them. When auxiliary cells and carpogonia form parts of the same branch-system, the aggregate is spoken of as a procarp. The threaas that establish contact with the auxiliary cells are spoken of as connecting filaments, as term that is especially appropriate in diverse Crytonemiales and Gigartinales, in whicn the auxiliary cells are widely separated from the carpogonia so that the connecting filaments are of considerable length. In the more specialised

Florideae (Ceramiales) however, the connecting filaments are extremely short, being represented merely by a small cell.

At the point of contact the membranes dissolve and communication is established between the cell of the connecting filament and the auxiliary cell. Among the less specialised types the same connecting filament may fuse successively with a number of auxiliary cells resulting in several centres of carpospore-formation, but in the higher Florideae there is fusion with only one auxiliary cell and, since there is often but a single connecting filament, only one group of gonimoblasts results from the act of sexual fusion.

In all Florideae, except Nemalionales, the carpospores are of course diploid, although they are produced from the gonimoblasts in essentially the same way as in haplobiontic forms. The carpospores give rise to a diploid plant, which usually resembles the haploid one in all respects, except that it normally bears only asexual sporangia forming four spores. The nuclear divisions in the sporangium are meiotic and the tetraspores grow into sexual individuals. The life cycle of these Florideae therefore includes a haploid sexual and two diploid asexual phases (the gonimoblasts producing carpospores the tetrasporic plant) and such Florideae are described as diplobiontic by Svedelius. There are three successive generations, although there is only one alternation of cytologicat phases in the life cycle. Lewis first demonstrated the isomorphic alternation between the sexual and the tetrasporic individuals experimentally (Figures 2.31 and 2.32)

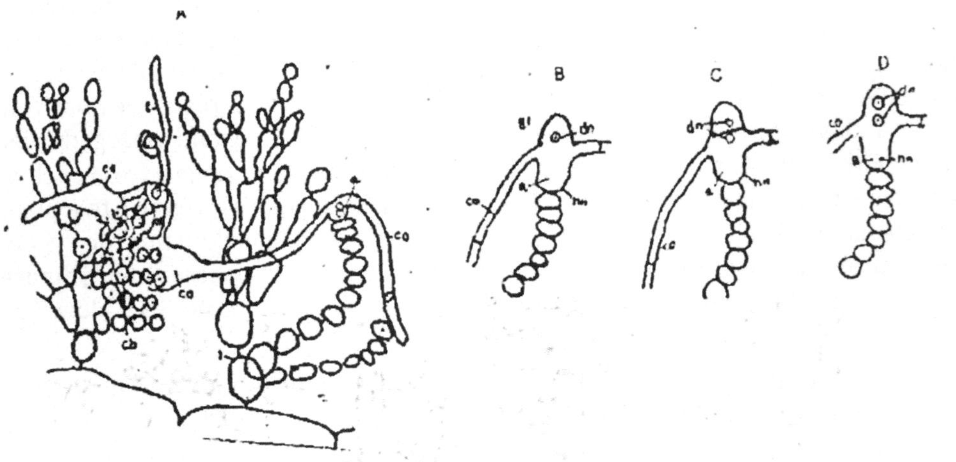

Figure 2.31: Post-fertilisation Changes in Diplobiontic Florideae (diagrammatic), A-D. *Acrosymphytum purpuriferum.*

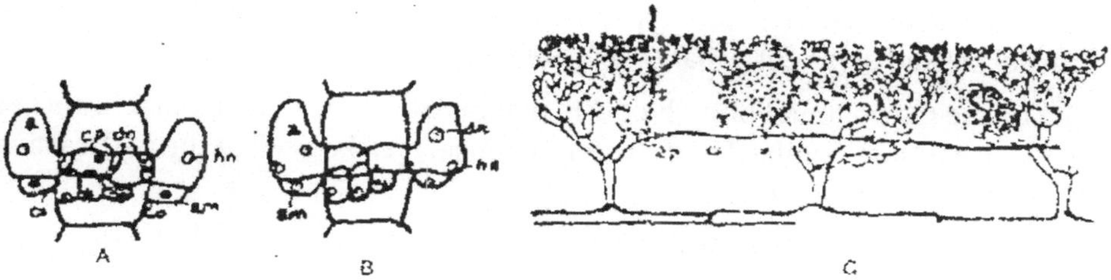

Figure 2.32: A-C: *Chondrus crispus;*
A, Habit; B, Transverse and C, Longitudinal section of mature thallus.

(e) Tetrasporangia and Tetraspores

The tetrasporangia of the diplobiontic Florideae are enlarged cells, which possess dense cytoplasmic contents and are almost invariably terminal on short lateral branches. This is clarly evident in such loosely branched forms as *Calithamnion* and *Antithamnion,* but even in genera with a compact structure, where they are embedded in the superficial tissues, their terminal position is generally recognisable without difficulty. The four spores either form a row (zonate) as in many Cryptonemiales and Gigartinales, or they are arranged cross wise (cruciate), commonly in the same plane, as for example in Rhodymeniales, or lastly they are grouped in a tetrad (tetrahedral) as in the majority of the Ceramiales. Different types may occur in different species of the same genus and occasional transitions are found. Certain Florideae especially Corallinaceae, produce only two spores (bispores) in the sporangia. Despite a few exceptions, the tetrasporangium of diplobiontic Florideae is usually the seat of meiosis.

Tetraspores and carpospores are nearly always heavier than water and tend to sink. The mature spores are nearly always filled with starch. During the early stages of germination as the reserves disappear, the chromatophores become apparent. There is no resting period. There are usually no essential differences in the germination of carpospores and tetraspores (Figure 2.33).

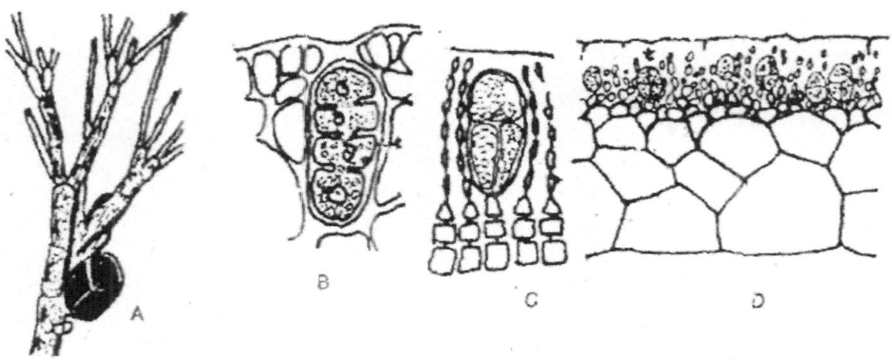

Figure 2.33: Tetrasporangia of Florideae: A, Cailithamnion corymbosum, (tetrahedral; B, Cystoclonium purpureum, (zonate); C, Peyssonneiia dubyi, (Cruciate); D, Chrysimenia ventrieosa, (Cruciate).

Reproduction and Triphasic type of life history (or) post fertilization changes in Gracilaria (Red alga) details are mentioned below (Figure 2.34).

- ☆ The Carpogonial branch of this regions of fronds Carpogonial branches are first recognized as a cluster of densely packed cell in the cortical and subcortical regions.
- ☆ Squash preparation of reproductive structure revealed 3-5 celled Carpogonial branch.
- ☆ The Carpogonium consist of a basel swellen portion and a short terminal trichogyne.
- ☆ The supporting cell also bears two celled auxiliary branch systems.
- ☆ Thus resulting a procap, frequently a few procarp occur adjacent to each other.
- ☆ Fertilized Carpogonium presumably fuses directly to the auxiliary cell. Then the auxiliary cell produce 3-5 celled gonimoblast.
- ☆ The cortical cells of the female gametophyte nearer to the developing Carposporophyte also divide to keep pace with the division of gonimoblast cells forming a dome shaped pericarp.
- ☆ The tissue below the pericarp anastomose and form the nutritive tissue.
- ☆ The terminal cells of gonimoblast filaments produce many cluster of Carposporangia. Each Carposporangium produce a single Carposore – Carpospores are spherical in shape. Mature Carpospores are released through the ostiole.
- ☆ Carpospores are germinate [by passive movements in the water currents] and give rise to tetrasporophytic plant.

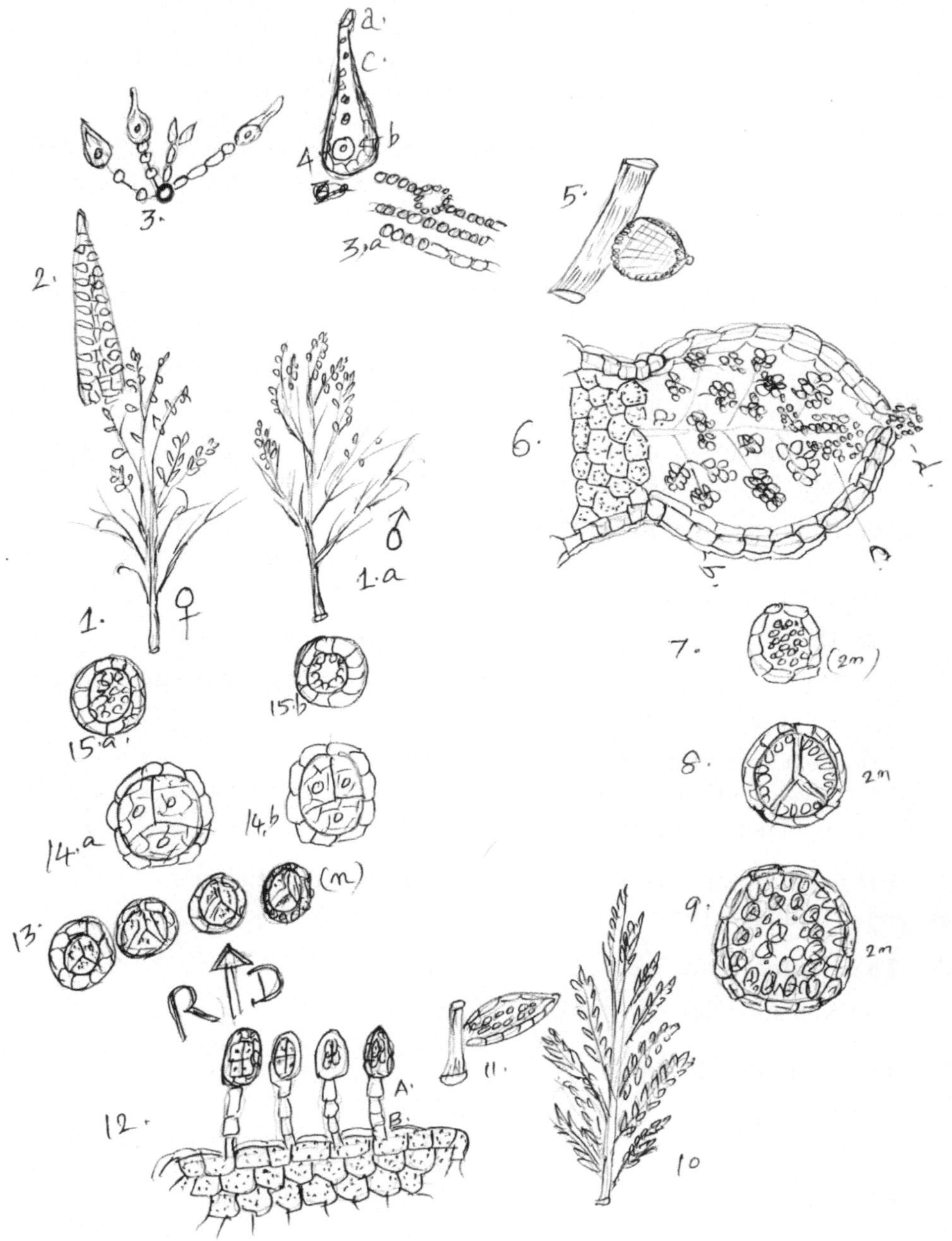

Figure 2.34: Triphasic Type of Life History (or) Post-fertilization Changes in Red Alga.

1) 1. Female gametophytic plant. 1a. Male gametophytic plant; 2) Mature female plant at the tip of clusters of Carpogonium; 3) Three celled Carpogonial branch (or) procarp. 3a. Spermatium; 4) Female nucleus seems frequently to move a short distance upwards to meet the male (Spermatium); 4a) Contact of a Spermatium with a trichogyne; 4b) Fertilization;

4c) Carpogonium is a flask shaped structure having a long necklike trichogyne; 5) Cystocarp; 6) a) Cystocap shows the development gonimoblast from the auxiliary cell, b) pericarp, c) mature Carpospore, d) releasing Carpospore by passive movements in the waterbody; 7) Carpospore (2n); 8) Carpospore germinate and give rise to Tetraspore (Cruciate form); 9) Tetrasporophyte Discoid microthallus; 10) Tetrasporophytic plant; 11) Short lateral branch lets or Ramuli; 12) Cortical cells are converted into Tetrasporangial initial. This initial divides into the (A) Upper Tetrasporangium and (B) Lower stalk cells; 13) Tetraspores (n); 14) a and b - Tetrasporellings; 15) a and b - Discoid microthallus.

- ☆ Tetrasporophytic plant posses tetrasporangia are distinct and present as sori.
- ☆ They are reddish round and oval swelling patches and encircle the basal portion of the short determinate branch lets or ramuli.
- ☆ Cortical cells are converted into tetrasporangial initials. This initial divides into the upper tetrasporangium and lower stalk.
- ☆ It enlarges and undergoes a transverse division and gives a binucleate condition. Then each content undergoes a further division resulting in a Tetrahedral arrangement.
- ☆ They are separated by means of simultaneous furrowing.
- ☆ Mature tetraspores are released by the dissolution of the upper portion of the sporangial wall.
- ☆ Each sorus initially occupies a distinct region if more number of sori develop in the same area the ramulus may became almost completely filled with tetrasporangia.

The following are the details about the haplobiontic and diplobiontic types as seen in Rhodophyceae:

A. The Haplobiontic Types

1. Nemalionales - *Liagora*

B. The Diplobiontic Types

1. Gelidiales - *Gelidium*
2. Cryptonemiales - *Amphiroa*
3. Gigartinales - *Gracilaria* and *Hypnea*
4. Rhodymeniales - *Champia*
5. Ceramiales - *Ceramium*

C. Rhodophyceae: Reproduction (Step-wise process)

1. Male organs	1. Female organs
2. Antheridium (Spermatium)	2. Carpogonium (ovum)

3. Passive movement (Zygote)
4. Mitotic division Germination Zygote as seen in Florideae except Nemalionales.
5. Gonimoblast filaments produced diploid.
6. A sexual spores (carpospores diploid)
7. Auxiliary cells+carpogonia (Procarp)
8. Carpospores germinate into a diploid asexual plant and produce asexual sporangia forming four spores (Tetrasporangia tetraspores)
9. Haploid sexual and two diploid asexual phases (the gonimoblasts producing carpospores, the tetrasporic plant).

3. Stable (Fertilisation)
4. Meiotic division Germination of of Zygote as seen in Nemalionales.
5. Gonimoblast filaments produced haploid.
6. Asexual spores (carpospores haploid).
7. Auxiliary cells + carpogonia (Procarp).
8. Carpospores germinate into a haploid sexual plant and produce sexual organs bears gametes.
9. Two haploid sexual and diploid asexual (zygote) stage.

Among the Rhodophyta, particularly the overwhelmingly large group, Florideae, we encounter a distinctively different kind of life history from any of the which is unique in being dependent on the female gametophyte, is introduced. Two common types of alternation of generations occur, involving this carposporophyte a diphasic type and a triphasic type.

Nemalion multifidum is one of the known of the diphasic types and consists of the alternation of a haploid gametophyte plant and a haploid carposporophyte borne upon it.

The triphasic type of life history is much more widespread in the Florideae and constitutes the most commonly observed life-cycle phenomenon of the red algae. *Gracilaria* serves as an example of this type. It shows an alternation of isomorphic, haploid gametophytes with a diploid carposporophyte and a diploid tetrasporophyte. The plant is a diplohaplont (diplobiontic).

Salient Features of *Polysiphonia*

1. The polysiphonous nature of the thallus which consists of an axial siphon surrounded by a number of pericentral siphons.
2. The plant body consisting of the decumbent portion and the upright portion.
3. The feathery, upright portion consisting of a much branched system of relatively large branches bearing numerous small branches called the trichoblasts. The latter bear the sex organs.

4. Growth in length by means of a single, dome-shaped apical cell.
5. Occurrence of pit connections between the adjoining central cells and between the central cells and the pericentral cells, another characteristic feature of the red algae.
6. Occurrence of three kinds of plants - gametophytes, carposporophyte and tetrasporophyte.
7. Sexual reproduction is oogamous. The male sex organs are called the spermatangia and the female carpogonia.
8. The spermatangia are developed in large numbers on the male trichoblast. The male gamete or the spermatium is non-motile.
9. The carpogonium is a flask-shaped structure consisting of the basal swollen portion and the distal tubular portion-the trichogyne.
10. The carpogonium is situated at the summit of a short, lateral, curved branch consisting of three or four cells. It is the carpogonium filament or procarp. The latter arises from one of the pericentral cells of the reduced female trichoblast.
11. The diploid nucleus formed by fertilization lies at the basal swollen portion of the carpogonium. The latter communicates by means of a tubular connection with the auxiliary cell cut off from the supporting cell.
12. The migration of one of the diploid nuclei into the auxiliary cell.
13. The development of gonimoblast filaments form the auxiliary cell.
14. The formation of a diploid carpospore in the modified terminal cell of the gonimoblast filament now called the carposrangium.
15. The development of the carposporophyte bearing the carpospores.
16. The formation of a fruit body or fructification called the cystocarp consisting of an urn-shaped sheath containing the carpospores.
17. Development of the tetrasporophyte by the germination of the diploid carpospore. The tetrasporophyte is concerned with production of meiospores called tetraspores.
18. Formation of 4 tetraspores in each tetrasporangium borne on the tetrasporophyte.
19. Meiosis occurs at the time of the differentiation of tetraspores which are thus haploid.
20. Germination of haploid tetraspores to give rise to the gametophyte or sexual plants concerned with sexual reproduction.
21. Distinct alternation Of generations consisting of morphologically alike sexual and asexual plants.

SELF LEARNING MATERIALS
(Red Algae)

1. Introduction

This class is both structurally and biochemically district from all other algal classes.

- ☆ Morphologially it is rather varied with unicells, filamentous, paranchymatous plants and diverse forms of pseudoparenchymatous organization.
- ☆ There is some differentiation of cells, particularly in the rather complex reproductive organs.
- ☆ Basic cell structure is simple with one or many plastids which have almost unique single thylakoid lamellae are nucleus and a number of mitochondria and Golgi bodies.
- ☆ Grains of the polysaccharide reserve material floridean starch, lie freely in the cytoplasm.
- ☆ No flagella are found in this group although a recent study (Simon-Bichard, Breadu, 1971-72) has shown the presence of structures, which are thought to be a pair of flagella, in male gametocyst of Bonnemaisonia.

2. Objectives

Eucaryotic cell, pigments chlorphyll a and d α-carotens phycoerythrin – phycocyanin, storage food floridean starch – cell walls of cellulose and pectin flagella entirely absent. Growth of the thallus may be apical as well as intercalary one of the striking features of red algae is the conspicuous cytoplasmic connection from cell to cell. Oogamous sexual reproduction is of general occurrence in red algae. The occurrence of non-aggregated photosynthetic lamella or thylakoids with the plastids.

3. Content

Majority of the forms are marine and only few are fresh water forms, usually red or dark in colour due to presence of accessory pigments, c-phycocyanin and r-phycoerythrin. Reserve food material in 'floridean starch', thallus in organised and exhibits complex differentiation, plasmodesmata is present between the cells except in the members of protoflorideae, sexual reproduction is specialized and oogamous motile cells altogether absent in the class and even the male sex cells are non-motile structures.

4. Revision Points

- ☆ Class Rhodophyceae is broadly classified into two subclasses namely Bangioideae and Florideae based on presence as well as absence of pitconnections.

- ☆ In Bangloideae thallus simple unicellular or multicellular whrease in Florideae, thallus relatively complex, uniaxial or polyaxial.
- ☆ Cell always uninucleate and growth through diffused intercalary cells in Bangiodeae whereas in Florideae cells multinucleate – Growth through a single apical cell.
- ☆ In Bangioideae each cell has a single centrally located stellate chromatophore whereas in florideae. Each cell has many discoid chromatophores, present along the periphery.
- ☆ Eucaryoticell, pigments chlorophyll a and d, α- and β-carotenes – phycoerythrin – phycocyanin – storage food floridean starch – cell walls of cellulose and pectin flagella entirely absent – non aggregated with single stranded thylakoids.

5. In Text Questions

1. Describe the triphasic diplobeoritic types of life cycles of an alga studied by you.
2. Give an account of life history of Gracilaria with necessary diagrams and point out the special features in it.
3. Describe the process of cystocarp formation in Gsraciharia.
4. Explain the changes that takes place after male and female gametes fuse in Gracilaria. Illustrate your answer with labeled diagrams.

5.. Agar agar is obtained from

a) Ectocapus	b) Fucus
c) Caulerpa	d) Gracilaria

6. The tetraspore germinat to produce

a) Tetrasporophyte	b) Sporophyte
c) Carposporophyte	d) Gametophyte

7. The sterile sheath of the Carposporophyte in Gracilaia is called

a) Epicarp	b) Procarp
c) Pericarp	d) Cystocarp

6. Summary

- ☆ The thallus of red algae is characterized by the presence of gelatinous matrial.
- ☆ The thalli of red algae are composed of highly organized cells whose wall is again composed of a mixture of cellulose and pectic material.

- Cells may be uni- or multinucleate. Each cell bears one to many chromatophores of varied shape.
- One of the striking features of most red algae is the presence of a pore-like opening in the cell wall through which conspicuous, broad cytoplasm connection is maintained from cell to cell known as pit connection.
- The chromatophores contain chlorophylls a and d. α- and β- carotenes, lutein, taraxanthin, allophycocyanin, a special red water soluble pigment r-pychoerythrin, and a blue pigment r-phycocyanin.
- The food reserves may be alcohols, oils, but the most important one is Floridean starch, an in soluble carbohydrates, a compound intermediate between the starch and dextrin.
- One of the outstanding characteristics of red algae is complete absence of flagellate reproductive structures.
- The male reproductive organ – Spermatangium is very simple structure which bears a single uninucleate single-celled male gamate – spermatium whereas the female are Carpogonium has a long neck-like structure – Trichogyne and a bulbous base – the carpogonial base where remains the female nucleus.
- The sporophyte bears tetrasporangia and tetraspores. Reduction division takes place during the sporogenesis in the sporanogium of the independent sporophyte. Such forms exhibit isomorphic to heteromorphic alternation of generations.

7. Supplementary Materials

1) Journal of Phycology,
2) Phycologia,
3) Journal of Applied Phycotogy,
4) Botanica Marina,
5) Phykos,
6) Indian Journal of Marine Science,
7) Seaweed Reseach and Utilization.

8. Assignments

- Structure and reproduction of subclass Bangioideae
- Describe the reproduction in red alga studied by you.
- With labeled diagrams, describe the structure of thallus and sexual reproduction in Gracilaria

9. Suggested Reading/References Books

1. Biology of Red algae, by Cole and Sheath: 1990.
2. Botany for degree students, Cambridge University Press, London. Part-I Algae by B.R.Vashista – 2006 by S. Chand and Company Ltd., Ram Nagar, New Delhi – 110055.
3. Introduction to the Algae; Structure and Reproduction, Herold C. Bold and Michael J. Wynne, 1978.

10. Learning Activities

☆ To understand the post-fertilization changes in the red algae.

11. Key Words

Polysaccharide, phycobilisomes, Floridearch starch, Single Stranded thylakoid, Pitconnection acontophora.

CYANOPHYCEAE

General Characteristic Features of Myxophyceae (Cyanophyceae)

The members of this class are commonly known as Blue green algae.recently this group is also called as cyanobacterium because of characteristics similarities with bacteria

1. The chief pigments of this are C-phycocyanin and C-phycoerythrin besides other usual pigments' (*e.g.* chl-a, b-carotene, flavacin, *etc.,* have the members are called blue green algae
2. The cell is prokaryotic in nature; means no definite nucleus, chromosome and subcellular organelles but nuclear material is present in the form of DNA.
3. Photosynthetic reserve food substance is cyanophycin starch.
4. Cell wall is made up of mucopeptide or muramic acid.
5. Vegetative reproduction by means fission, fragmentation, hormogonia, akinetes, endospores, exospores and nanocyst, sexual reproduction is completely absent.

Classification

Cyanophyceae

Tribe-A

Coccoid

- ☆ Chrococcales *e.g.*, Gloeocapsa
- ☆ Aphanacapsa, Gloeatheca
- ☆ Chemosiphonales

Tribe-B

Hormogonia

- ☆ Oscillatoriales *e.g.*, Oscilatoria
- ☆ Nostocales *e.g.*, Nostoc
- ☆ Scytonematales *e.g.*, Scytonem
- ☆ Stigonematales *e.g.*, Stigonema
- ☆ Rivulariales *e.g.*, Rivularia

MICROCYSTIS

Systematic Position

Division	: Cyanophycophyta
Class	: Myxophyceae (or) Cyanophyceae
Order	: Chroococales
Family	: Chroococaceae
Genus	: *Microcystis*

Characteristic Features

1. *Microcystis* is a fresh water genus; they often cause water blooms, in hard water lakes.
2. The thallus is free-floating and varies in shape.
3. It contains a mass of single spherical cells but the sheaths of individual cells are confluent with colonial envelope.
4. The gelatanious matrix of the colonial envelope is of watery consistency and the margins of the matrix are usually not evident.
5. The cells frequently contain numerous pseudovacuoles
6. The reproduction of individual cells takes place by means of fission in three planes.
7. The reproduction of the colony is through successive disintegration, each portion develops into a new colony.
8. The shape of the colony is primarily determined by the environmental conditions.

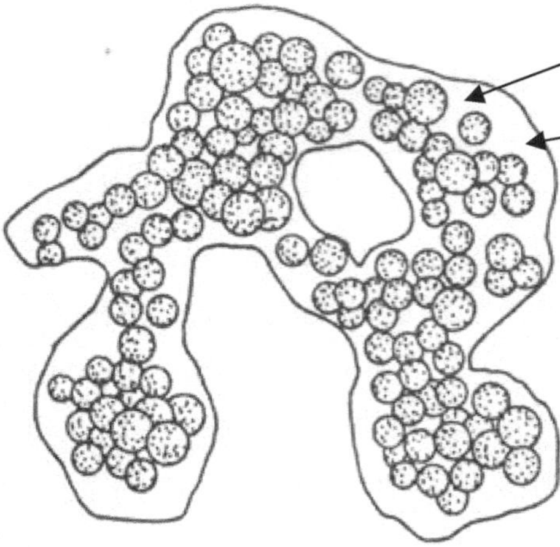

Figure 2.35

Oscillatoria

Systematic Position

Division	: Cyanophycophyta
Class	: Myxophyceae (or) Cyanophyceae
Order	: Oscillatoriales
Family	: Oscillatoriaceae
Genus	: *Oscillatoria*

Characteristic Features

1. It exhibits a very simple type of construction having unbranched tri-chomes.
2. The trichome consists of a row of identical cells, and is surrounded by a sheath of mucilage.
3. All the cells are capable of division and growth. In some species, the terminal cells are conical or rounded, often showing attenuation into capitate structure.

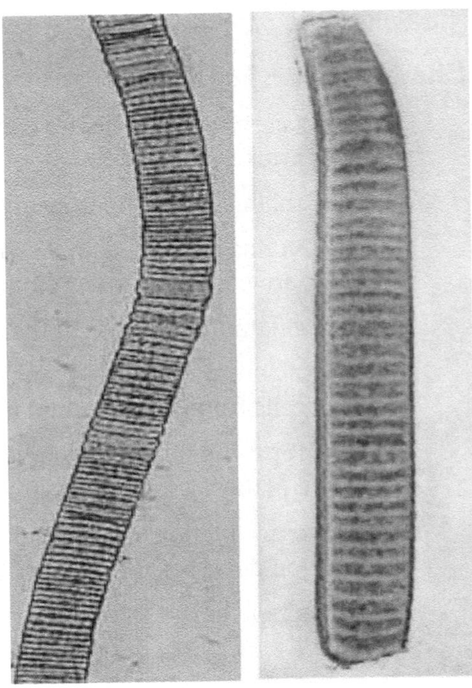

Figure 2.36

4. The intersurface of the terminal cell of the trichome may be covered by a thickened cap (calyptra) which is protective in function.

5. The reserve foods are cyanophycin granules. and cyanophycean starch. They are found in the peripheral portion of cytoplasm.

NOSTOC

Systematic Position

Division	: Cyanophycophyta
Class	: Myxophyceae (or) Cyanophyceae
Order	: Nostocales
Family	: Nostacaceae
Genus	: *Nostoc*

Characteristic Features

1. *Nostoc* is a common blue green alga of filamentous form
2. Species of *Nostoc* may be terrestrial or aquatic and generally occur in ponds, ditches and other pools of water as well as damp soil in the form of small, firm mosses of jelly.
3. A few species are endophytic in habit occurring in the intercellular cavities of plants like *Anthoceras,* Lemna and roots *Cycas.*
4. Some leads a symbiotic life in association with a fungus farming lichen.
5. Each gelatinous mass contains numerous unbranched filaments, which look very much like strings of beads under a microscope.
6. Each filament has a gelatinous sheath of its own, in addition to the gelatinous matrix of its colony.
7. The filament minus the sheath is commonly called a trichome.
8. Each cell of the filament is blue-green, more or less spherical or oval.
9. The characteristic feature of *Nostoc* is the presence of some enlarged vegetative cells, terminal or intercalary, with thickened walls and transparent content called heterocysts.
10. At the two poles of each heterocyst there are two pores, one at each end, through which cytoplasmic connection is maintained with adjacent cells.
11. The reproduction is vegetative and takes place by means of harmogones akinetes and rarely heterocysts.
12. In favourable conditions the filament break into small pieces called hormogone consisted of two or more cells capable to give rise to a new plant.
13. During unfavourable condition the vegetative cells become thick walled, containing reserve food called akinetes or resting spore, which give rise to new filament during favorable condition.

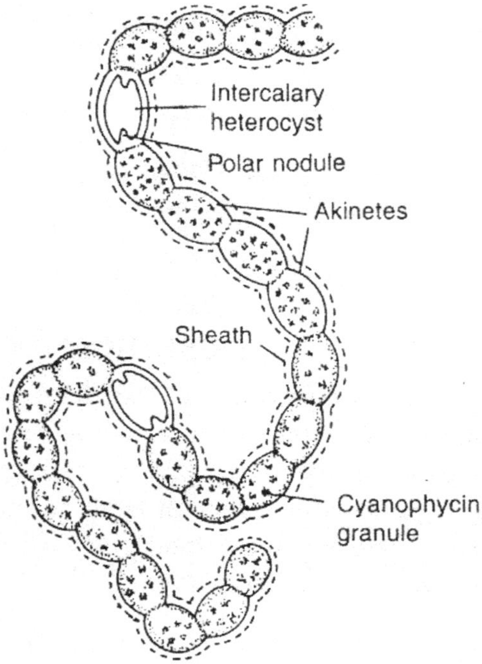

a) *Nostoc* filament

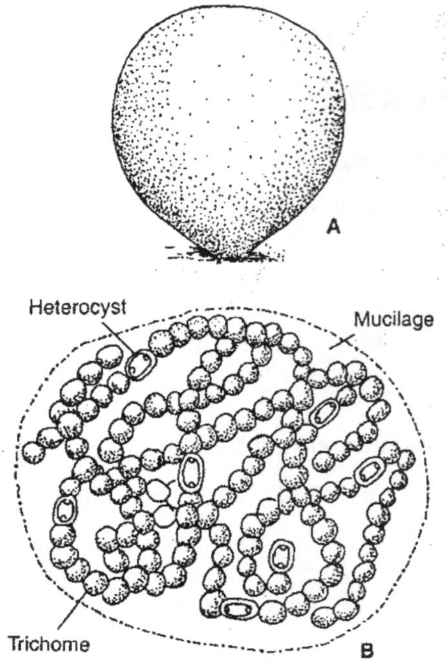

b) *Nostoc* colony

Figure 2.37

Spirulina

Systematic Position

Division	: Cyanophycophyta
Class	: Cyanophyceae
Order	: Nostocales
Family	: Oscillatariaceae
Genus	: *Spirulina*

Characteristic Features

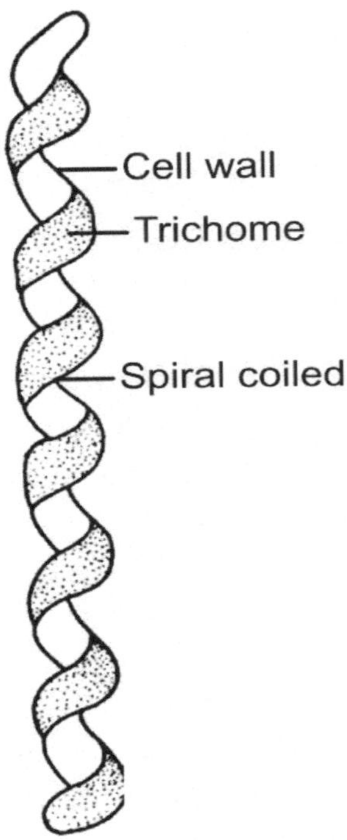

Figure 2.38

1. *Spirulina* is multicellular freshwater blue green algae commonly found in various polluted and unpolluted ponds.
2. The genus *Spirulina* contains the group of filamentous cyanobacteria characterized by spiral shaped chains of cell.
3. *Spirulina* showed trichome deep blue green, spirals regular, 4-6 number per trichome, at the end conical, attenuated.

4. Trichome 3 μm broad, 14.4 μm long, distance between two spiral being 31-33 μm.
5. The breadth of spirals 27 μm cells indistinct seta not clear.

Anabaena

Systematic Position

Division	: Cyanophycophyta
Class	: Myxophyceae (or) Cyanophyceae
Order	: Nostocales
Family	: Nostacaceae
Genus	: *Anabaena*

Characteristic Features

1. *Anabaena* is widely distributed and is commonly found in fresh water, floating freely or in thin mucous layers, and also after in wet soil rocks and tree trunks.
2. The plant body is unbranched, filamentous and consists of a row of oval or spherical cells. Bead like in appearance, intermixed at frequent intervals with some heterocysts and akinites.

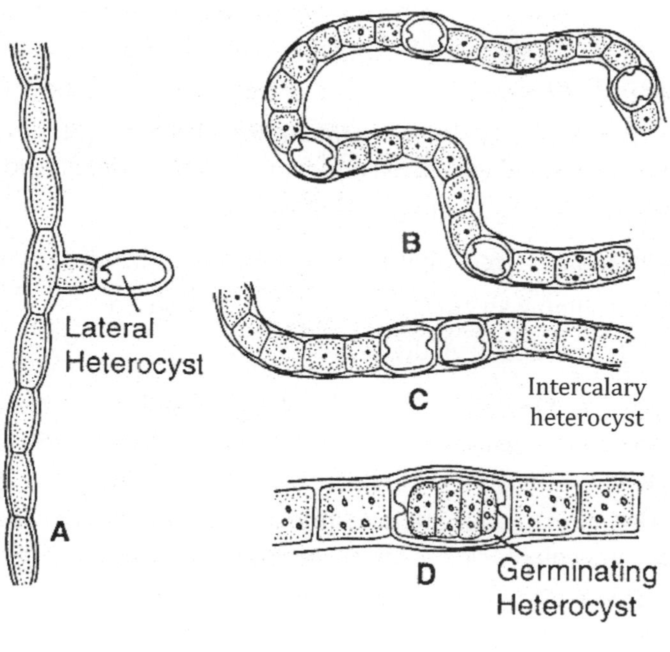

Figure 2.39

3. The sheath which surrounds the trichomes are hyaline and generally of watery nature.
4. The cells are usually, spherical or barrel-shaped, rarely cylindrical and never discoid.
5. The protoplast of the vegetative cell are usually filled with numerous *Pseudo vacuoles.*
6. The intercallery heterocyst are of the same shape as vegetative cells, though slightly larger are present in any trichome.
7. The reproduction is vegetative and takes place by means of hormogones, akinetes and rarely heterocyst.
8. Akinetes develop only next to heterocyst, always larger than the vegetative cells and generally cylindrical and with rounded ends.

Lyngbya

Systematic Position

Division	: Cyanophycophyta
Class	: Myxophyceae (or) Cyanophyceae
Order	: Nostocales
Family	: Oscillatoriaceae
Genus	: *Lyngbya*

Characteristic Features

1. The species of *Lyngbya* are commonly found in freshwater such as ditches, ponds, pools and even in drains some in damp soil and rocks.
2. The plant body is uniseriate, unbranched filament.
3. The trichomes occur singly in the enveloping sheath.
4. The enveloping sheath is often thick and coloured.
5. All cells are capable of division
6. The members do not form spores or heterocysts
7. The reproduction takes place by vegetative method by means of hormogones.
8. The filament breaks into two or more celled hormogones each develops into a new plant.

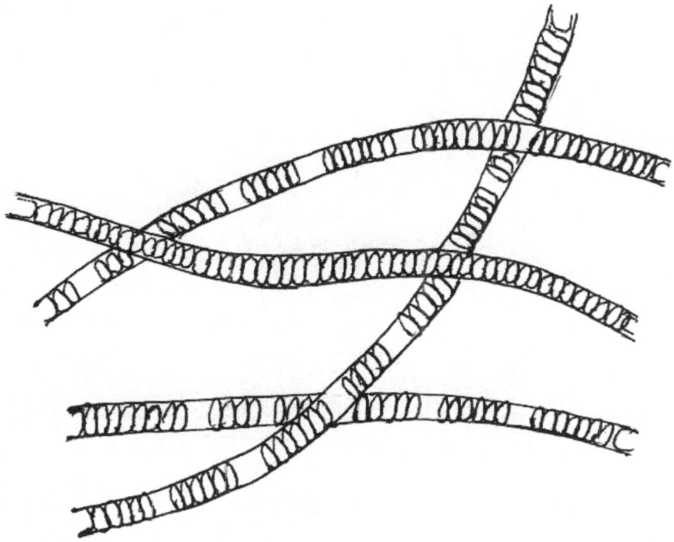

A portion enlarged view

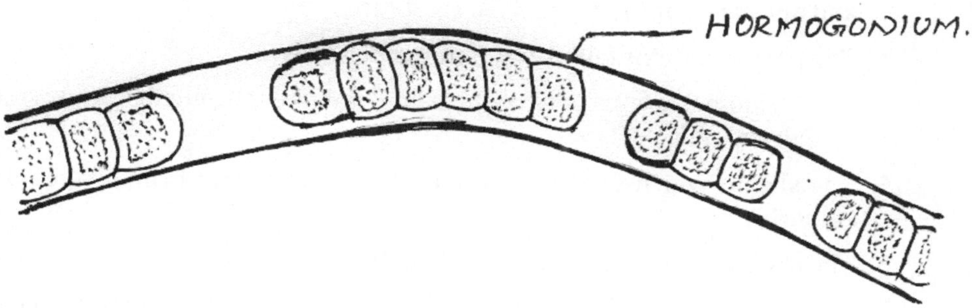

Figure 2.40

Scytonema

Systematic Position

Division	: Cyanophycophyta
Class	: Myxophyceae (or) Cyanophyceae
Order	: Nostocales
Family	: Scytanemataceae
Genus	: *Scytonema*

Characteristic Features

1. The genus is usually found in subaeri al habitats best on damp soil and rocky cliff.

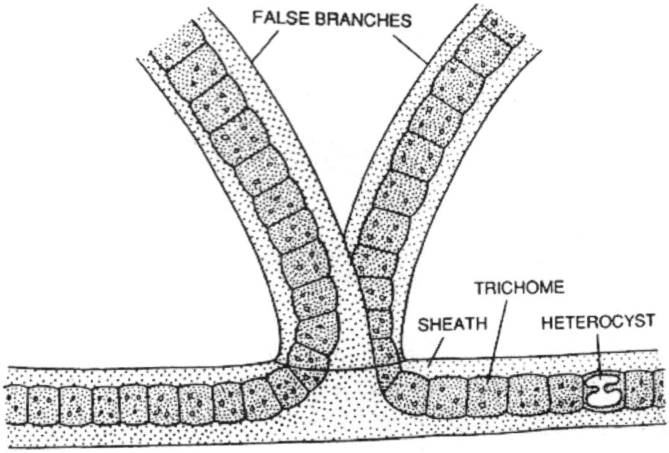

Figure 2.41

2. The filament that consist of trichome and sheath posses distinct basal and apical regions forming little erect tufts.
3. The genus contains false type of branching
4. The trichomes are usually of the same diameter throughout and with cylindrical cells
5. The sheath which surrounds the trichomes is always firm, hyaline or coloured
6. The sheaths are homogeneous or lamellated
7. The heterocysts are intercalary and are borne single or in twos or threes.
8. The heterocysts are of the same size as vegetative cell
9. Alkinetes are rare.

The genus contains false type of branching. The branches arise either between two heterocyst or else adjoining one as a result of the degeneration of an intercalary cell. The intercalary growth results in strong pressure being applied to the sheath, which finally ruptures and the trichome forms a loop outside. Further growth result in breaking of this loop and twin branches are produced. One or both of these branches may subsequently proceed to additional growth, the branch sheath extending back into parent sheath. Sometimes the false branching is initiated by degeneration of a heterocyst or a vegetative cell and subsequent growth of two filaments on either side.

Stigonema

Systematic Position

Division	: Cyanophycophyta
Class	: Myxophyceae (or) Cyanophyceae
Order	: Stigonematales
Family	: Stigonemataceae
Genus	: *Stigonema*

Characteristic Features

1. In Stigonema the threads consist of a single row or multiple rows of cells with definite apical growth.
2. Cells, enclosed within a gelatinous sheath that varies in thickness.
3. Behind the apical cell, one of the products divides twice to give a cruciform group.
4. One of these cells gradually moves into the centre and by further division a central cell surrounded by four pericentrals is produced.
5. These pericentrals divide again radially so that finally successive tires of pericentrals are produced, arranged approximately longitudinally. These cells are connected by protoplasmic strands.
6. The filaments are often aggregated to form a macroscopic yellow-green gelatinous mass.
7. The species are most common where there is continual dripping of water.

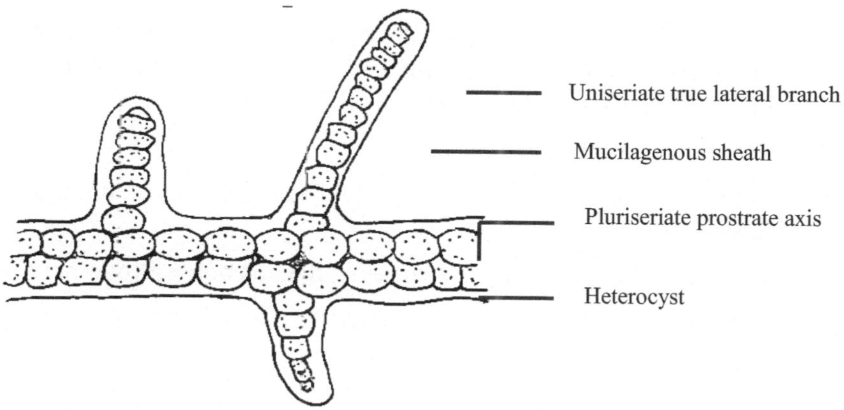

Figure 2.42

Distinguishing Features of Gyanophyceae

Some important distinguishing features of blue-green algae are as follows:

1. They are unicellular (*e.g., Chroococcus*), colonial (*e.g., Gloeocapsa, Gloeotrichia*) or form chains of cells - trichomes (*e.g., Oscillatoria, Scytonema*). In colonial forms, all cells of a colony occur in a common gelatinous matrix. The cells of the colony may divide to form definite (*e.g., Merismopedia*) or indefinite and irregular (*e.g., Aphanotheca*) assemblage of cells. In filamentous forms (a trichome surrounded by a sheath is called a filament), each trichome has its own gelatinous sheath. The filaments are usually unbranched and uniseriate (*e.g., Anabaena, Nostoc, Lyngbya*) or may sometimes show pseudo-branching (*e.g., Scytonema, Tolypothrix*).
2. Their cell structure is prokaryotic. They lack membrane bound organelles and the nucleus is of incipient type.
3. The cell wall is composed of microfibrils held together in a mucilage sheath. The orientation of fibres may vary in different species, and may be reticulate (*e.g., Nostoc*) or long intertwining parallel (*e.g., Lyngbya*). It helps in sloughing off parasites and in retaining water, thus avoiding desiccation (particularly in those species which do not produce resting spores).
4. The cell wall (external to plasmalemma) is differentiated into four layers. It is composed of mucopeptide, together with carbohydrates, amino acids and fatty acids.
5. The principal pigments are chlorophyll-. a, c-phycocyanin and c-phycoerythrin, and they impart characteristic blue-green colour to these algae.
6. They lack membrane bound plastids. Instead, pigments are found embedded within lamellae composed of two membranes joined at the ends. These structures are known as thylakoids.
7. They are essentially aerobic.
8. The prokaryotic cells of the blue-green algae do not contain membrane bound nucleus.
9. They show gliding or jerky movements; flagella being altogether absent.
10. The reserve food material is a special type of starch (cyanophycean starch) and protein (cyanophycin granules).
11. In some blue-green algae (*e.g., Nostoc, Scytonema*) specialised structures, known as heterocysts, are present. They may be terminal (*e.g., Gloeotrichia*) or intercalary (*e.g., Nostoc*) in position. Definite function of heterocysts is not known, but they are supposed to be the centre of N_2-fixation. Heterocyst-forming cyanophycean algae are multicellular in

which growth requires the activity of two metabolically interdependent cell types—the vegetative cells that perform oxygenic photosynthesis and the dinitrogen-fixing heterocysts. Vegetative cells provide the heterocysts with reduced carbon, and heterocysts provide the vegetative cells with fixed nitrogen.

12. Reproduction takes place by vegetative and asexual methods. Sex organs, gametes and flagellated zoospores are altogether absent. However, in some (*e.g., Anabaena, Nostoc, Synechococcus, Cylindrospermum*) genetic recombination has been reported.
13. Vegetative reproduction takes place by cell division, fragmentation and hormogonia formation.
14. A sexual reproduction takes place by akinetes, endospores, exospores, heterocysts and nannocytes.

Cell Structure

The blue-green algae have prokaryotic cell organization (Figure 2.42). The internal membranes which separate DNA from the cytoplasm and isolate enzymatic machinery of photosynthesis and respiration in specific organelles are absent. The cell as such is devoid of organized nuclei, chromatophores and mitochondria. The cell wall contains a specific mucopeptide as its strengthening component.

1. Sheath and Cell Wall

The presence of the mucilagenous sheath is a constant feature of blue-green algae. The sheath consists of three layers of microfibrils reticulately arranged within an amorphous matrix. This gives a homogeneous appearance to the sheath. The mucilagenous sheath retains absorbed water which is useful during the periods of desiccation and for sloughing off parasites. A double-layered cell wall is present inside the sheath. The inner layer of this wall contains mucopeptide and muramic acid. When viewed under electron microscope, the wall shows four layers, designated as LI, LII, LIII and LIV. Each layer is about 10 nm in thickness, and the LI lies next to plasmalemma. In structure and composition the cell wall of blue-green algae is similar to that of Gram-positive bacteria; the cell wall in both is composed of mucopeptide together with carbohydrates, amino acids and fatty acids. The LII layer has mainly peptidoglycan. The LI and LIII layers are electron transparent, whereas the LII and LIV layers are electron opaque.

A cytoplasmic membrane (plasmalemma) is present inner to the cell wall and this membrane consists of two electron opaque layers separated by a translucent layer. The cytoplasmic membrane invaginates inside the cell, and these invaginations are considered sites of biochemical functions normally associated with mitochondria, endoplasmic reticulum and Golgi bodies in eukaryotic cells.

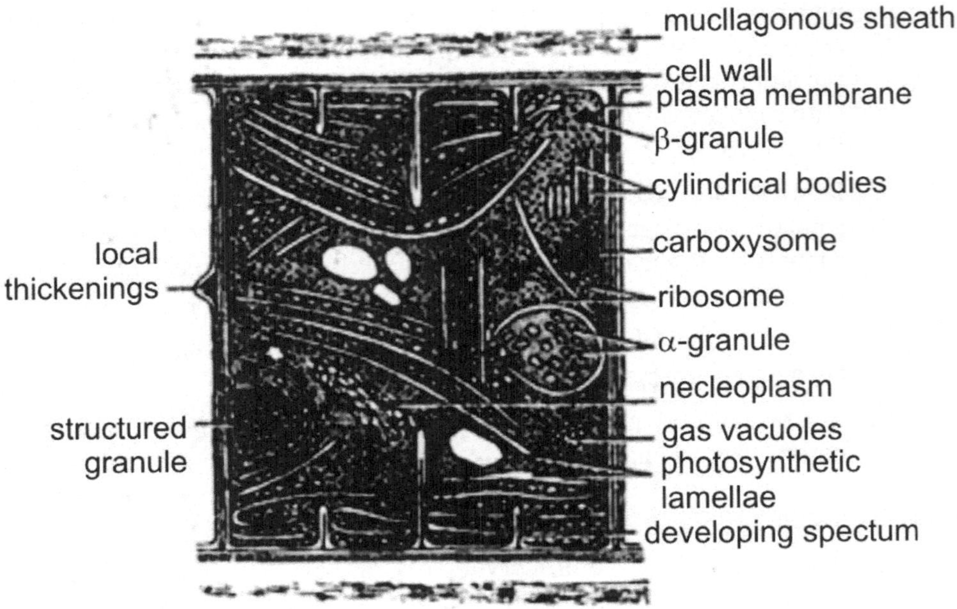

Figure 2.43: Ultrastructure of a Typical Cyanophycean Cell (*e.g., Nostoc*).

2. Photosynthetic Lamellae

Electron microscopic studies have revealed the presence of a complex lamellar system (thylakoid) which is functionally analogous to the plastids of eukaryotic cells. The lamellae in cyanophycean cell are not separated from the remainder of the cytoplasm in a membrane-bounded organelle. They appear as elongated, flattened sacs consisting of two unit membranes, each about 75 Å thick: Adjacent thylakoids are separated from each other by a space of 50 nm, occupied by contiguous rows of discoidal phycobilisomes. These phycobilisomes contain photosynthetic pigments, c-phycocyanin and c-phycoerythrin.

3. Intracytoplasmic Inclusions

Various kinds of subcellular organelles have been observed in cyanophycean cells and these include cyanophycin granules, polyhedral bodies, ribosomes, polyglucoside bodies, polyphosphate bodies, α-granules, β-granules and gas vacuoles.

The significance and functions of these organelles are not fully understood. Ribosomes are involved in protein synthesis, cyanophycin granules are regarded as reserve food, and polyhedral bodies (now called carboxysomes) contain an important enzyme, ribulose diphosphate carboxylase. Several planktonic species of blue-green algae contain numerous small gas vacuoles (pseudovacuoles). These vacuoles are not tonoplast bound and contain metabolic gases. They are as such a provision for buoyancy regulation. Under electron microscope, the gas vacuoles appear to consist of packed arrays of cylindrical vesicles. The gas vacuoles are produced in low light intensity and disappear in high light intensity. They refract light and as such cells appear purplish, brown or even black under microscope.

4. Nucleoplasm

The central region of a blue-green algal cell appears somewhat transparent and contains most of the genetic material. An organized nucleus with a nuclear membrane and nucleolus is absent. Since DNA is not associated with protein material (histones or protamines), organized chromosomes, characteristic of the eukaryotic cell, are not found in the cyanophycean cell. RNA is present in addition to DNA.

PhyloWgeny and Affinites of Cyanophyceae

There are many evidences which indicate that Cyanophyceae are the most primitive of algal organisms. Their prokaryotic cell structure, absence of division of labour and flagellate forms, and lack of sexuality are some of the characters which support their primitive nature. Fossil evidences indicate their origin in early Precambrian period.

Their prokaryotic cell structure and absence of flagella lead to believe that they originated from unicellular forms devoid of flagella. Since most of the members of Cyanophyceae are terrestrial, many investigators believe that they originated from terrestrial ancestors rather than aquatic.

Similarity of blue-green algae with red algae in certain characters, such as the presence of similar phycobilins and complete absence of flagellated reproductive cells suggest some affinity in these two groups. However, some other important characters of blue-greens, such as prokaryotic cell structure, and the presence of heterocysts, hormogonia and akinetes are not found in red algae. This indicates that there does not appear any close relationship between these two groups.

Since the recognition of blue-greens as a biological group in the early 19th century, they have been treated as a class or division of algae. But around I960, when fundamental differences in cellular organization between prokaryotes and eukaryotes were fully elucidated, it became evident that they are more closely related to the prokaryotic bacteria than to the eukaryotic algae. However, a detailed study of the nature of their photosynthetic pigments and capacity to perform oxygenic photosynthesis show that the only unifying character in Cyanophyceae and bacteria is prokaryotic structure of the cell.

The affinities of Cyanophyceae with Rhodophyceae and bacteria are summarised below.

[A] Similarities with Red Algae

1. Flagellated motile cells are absent in both groups.

2. The blue (c-phycocyanin) and red (c-phycoerythrin) pigments present in Cyanophyceae are chemically similar to those found in Rhodophyceae (r-phycocyanin and r-phycoerythrin).
3. Pit connections are present in the family Stigonemataceae of the class Cyanophyceae, and similar structures are also found in Rhodophyceae of the class Gracilairaceae.
4. The mechanism of fatty acid synthesis is similar in both groups.

[B] Dissimilarities with Red Algae

1. The blue-green algae show prokaryotic cell structure, whereas it is typical eukaryotic in the red algae.
2. Cells of the blue-greens are covered with a mucilagenous sheath, which is absent in the red algae.
3. Presence of hormogonia and heterocysts is a characteristic of the most blue-greens, but these structures are not found in the red algae.
4. Sexual reproduction is absent in the blue-green algae, whereas the red algae show an advanced type of sexual reproduction.

[C] Affinities with Bacteria

1. Both, bacteria and blue-greens show typical prokaryotic structure.
2. Like bacteria, the blue-green algae have an incipient nucleus (nuclear material without nuclear membrane), and lack membrane bound plastids.
3. The mucilagenous sheath surrounding blue-green cells and the capsule present in many bacteria have a similar structure, as both are made up of extremely fine fibrils.
4. Some bacteria (*e.g., Tolypothrix*) are structurally similar to hormogoriia, found in the blue-green algae.
5. *Beggiatoa,* a sulphur bacterium, resembles with *Oscillatoria* in shape and movement.
6. Like bacteria, the blue-green algae are also sensitive to antibiotics.
7. Many metabolic processes (*e.g.,* sulphur and nitrogen metabolism) are similar in both groups.
8. Recombination of genetic material, similar to bacteria, has been reported in some blue-greens (*e.g., Anacystis nidulans*).

SELF LEARNING MATERIALS

(Blue Green Algae)

1. Introduction

- ☆ It is a small primitive group comprising of about 2500 species placed under 150 genera. All of them are included in a single class cyanophyceae.
- ☆ The members of this class are considered the simplest, living autotropic plants the cells constituting the thallus are prokaryotic.
- ☆ The flagella are entirely lacking, when locomotion occurs, it is of glinding, Jerky type.
- ☆ The phycobilin pigments unique to this class are blue C. phycocyanin and C.phycocrythrin in addition to Chlorophyll-a, β-carotena and unique. Xnthophylls namely myxoxanthin and mycoxanthophyll.
- ☆ The unique food-storage compounds are mycophycean starch and a proteinacean material cyanophycin.
- ☆ Many filamentous blue-green possess specialized cells of disputed function known as the heterocysts.

2. Objectives

- ☆ Presence of Blue green pigmentation in the peripheral cytoplasm.
- ☆ They often copious production of mucilage, frequently in the form of well-defined sheaths to the cells or filaments.
- ☆ The absence of advanced stages of thallus organization.
- ☆ The product of photosynthesis is cyemophycean starch present in granular form.
- ☆ The propagation taking place by a simple division, by non-motile endospores, exospores, spores hormogonia, heterocyst or else by vegetative fragmentation.

3. Content

Blue green algae include unicellular to filamentous organism: prokaryotic cell, true nucleus and chromatophores lacking; pigmentation includes chlorophyll α, β carotene, various xanthophylls, and two C. phycobilins storage food Cyanophycean starch, flagellation absent reproduction by accidental breaking and by production of specialized cells, sexual system most inefficient.

4. Revision Points

1. Unicellular thick-sheathed plant body in Chroococcaceae Reproduction by fission.

2. In oscillationaceae plant body exhibits oscillating, bending or gliding movement. Filamentous unbranched plant body bearing extremlly thin sheath reproduction by accidental breaking and by separation disc.
3. In Nostocaceae filaments sheathed and are held together by soft mucilage, cells rounded, presence of heterocyst in the filament, reproduction by hormogonia.
4. In scytonemataceae filaments body with false branching presence of heterocyst in the filaments.
5. In Rivulaisaceae, Filamentous unbranched vegetative body is differentiated into base and apex presence of heterocyst at the broad basal end of the trichome, the filament is partly covered by a sheath.

5. Intext Question

1. Describe the range of thallus structure in Cyanophyceae.
2. Describe the different modes of reproduction met with in Cyanophyceae.
3. Discuss the phylogeny and inter relationships of myxcophyceae.
4. Write an essay an economic importance of Blue-green algae.

6. Summary

Habitat: Aquatic or in moist situation pigments; Chlorophyll α, β-Carotene billi protein, Myocoxanthin and mycoxanthophyll, food reserve, glycogen.

Cyenophycean: Cell well components, Hemicellulose, Cellulose, possibly mucopeptide. Reproduction a sexual genetic recombination observed but it mechanism unknown. Growth form, unicellular, cellular aggregate filaments, Flagella none.

7. Terminal Exercises

1. Water blooms are formed due to the presence of

   a) Hydrilla — b) Bacteria

   c) Blue green algae — d) Green algae

2. Plastids are absent in the members of class

   a) Chlorophyceae — b) Rhodophyce

   c) Myxo phyceae — d) Phaeophyee

3. Which of the following has ability to fix atmospheric nitrogen

   a) Red algae — b) Blue green algae

   c) Brown algae — d) Chlorella

8. Same as in Rhodophceae

The structure and reproduction of the Algae by F.E.Fritsch 1945, Cambridge University Press, London.

9. Assignments

1. Discuss habitat, structure and reproduction of Cyanophycean member.
2. Illustrate the structure of a cell of a cyanophycea algae under electron microscope.
3. "Heterocyst is popularly called Botanical Engima". Discuss the statement in the light its structure and function.

10. Learning Activities

1. List the features which indicate that the blue green algae are primitive and ancient members of the plant kingdom. In what respects do they resembly bacteria?
2. To understand the cell type of bacteria and blue green algae? Give an account of Morphological features, physiological functions and methods of reproduction in any of the type studied by you.

11. Keywords

Prokaryotic cell, cynophycean starch. Absence of Motile cells, heterocyst unicellular, colonial form and filamentous plant body.

Unit 3

Origin and Evolution of Sex in Algae

There are two primary methods of reproduction among plants, asexual and sexual. In asexual reproduction no sex is involved. There is no union of elements of two cells. This type of reproduction is brought about by the formation of special cells set apart for this purpose. Each of these cells is capable of developing into a plant like the parent on separation. Such special cells are called the asexual spores. Sexual reproduction, on the other hand, involves the union of two special cells called the gametes. The two gametes fuse to form the zygote which develops into a new individual. Normally the gametes are incapable of growth alone.

(a) Origin of Sex

It involves the origin of gametes. The most primitive algae, the Cyanophyceae reproduce only by the asexual methods of reproduction. *Protococcus* reproduces entirely by fission. Sexual reproduction is entirely absent in their case. Other algal forms whether unicellular or multicellular reproduce both by asexual as well as sexual modes of reproduction. With a few exceptions sexual or gametic reproduction has not replaced asexual reproduction, but has been added as a supplementary method. It is evident that originally the gametes were derived from the motile asexual spores of zoospores which had become too small and thus weak to form a new plant alone. Except for size the gametes are much like the zoospores.

The view is supported by the fact that in algae like *Chlamydomonas, Ulothrix* and many others with isogamous mode of sexual reproduction, the motile asexual spores (zoospores) and the gametes intergrade. In forms like *Chlamydomonas debaryanum,* the *zoospores* and the *gametes* are identical in every respect. They are similar in form, structure and mode of development. The chief morphological difference between them is their size. The zoospores are larger and the gametes are smaller. This difference in size depends upon the number of divisions that occur at the time of their differentiation. Where divisions stop early zoospores which are comparatively larger in size are formed. They contain the minimum level of factors essential to enable them to grow into new individuals alone. When division is long continued it results in the formation of smaller, swarmers resembling zoospores. They are incapable of giving rise to new plants alone. These come to be known as the gametes or sexual cells.

In *Ulothrix zonata* three kinds of zoospores have been reported :

(*a*) Quadriflagellated macrozoospores formed in small numbers per cell;

(*b*) Quadriflagellated microzoospores formed in large number per cell; and

(c) Biflagellated microzoospores formed in still large numbers per cell.

The gametes in *Ulothrix* are also biflagellated and are produced in large numbers. These gametes closely resemble the zoospores but are smaller than the macrozoospores and quadriflagellated microzoospores in size, and have two flagella instead of four. In structure, both the gametes and zoospores are similar and produced in the same manner. The difference in size and number of flagella between the gametes and zoospores of *Ulothorix* has been bridged over by the reported occurrence of biflagellated microzoospores in *U. zonata.* So the principal difference between the gametes and zoospores, as mentioned above, lies only in the number of divisions occurring at the time of their differentiation. The similarity of gametes to zoospores suggests that sex (union of two cells) has originated by the accidental fusion of zoospores which had become too small to form a new plant alone. The fusion probably took place in response to certain unfavourable environmental conditions. Since this accidental pairing proved of advantage to the plant in increased vitality and vigour, the method was maintained and finally became fixed.

This view is supported by the fact that sexual reproduction in the lower organisms is always associated with conditions unfavourable for growth. So long conditions for growth are exceptionally favourable they multiply by vegetative means. Under favourable conditions of growth they propagate by spore formation. Sexual reproduction is resorted to only under conditions unfavourable for growth. This is due to the fact that in enfeebled conditions caused by starvation, the plants produce small swarmers which along are unable to produce new plants. In pairing together they combine their resources and the resultant fusion cell or the zygote regains the power of producing a new individual directly. The fact that the zygote normally functions as a thick-walled resting spore in green algae suggests that sexual reproduction in green algae originated as an additional method to enable them to tide over a period unfavourable for growth. With the return of favourable conditions for growth the zygote in many green algae germinates to form asexual spores. It is merely a response to the favourable conditions.

The very fact that under certain conditions the gametes behave like spores and develop into new plants without fusion lends additional support to the view that the gametes are modified spores.

Evolution of Sex

Originally when sexual reproduction had become established both the gametes were flagellated and of the same size. Morphologically as well as physiologically they were identical. This condition is retained by some of the green and brown algae such as *Chlamydomonas debaryanum, Cladophora* and *Ectocarpus.* In all these forms sexual reproduction is seen in its simplest form. It consists in the fusion of similar, motile and naked gametes. They are called the isogametes. They fuse in water. The isogametes are indistinguishable as to sex and are produced in ordinary vegetative (undifferentiated) cells. This kind of sexual reproduction which consists in the fusion of morphologically and physiologically similar gametes is called isogamy. It is the most primitive method of sexual reproduction. It marks the transition from asexual reproduction to sexual reproduction.

There are some forms among the algae which represent all the steps through which plants passed in the evolution of sexual differentiation. In some green algae the fusing gametes are externally alike but dissimilar in their behaviour. In *Ulothrix,* for example, the gametes are similar but fusion takes place only between two gametes produced in distinct filaments. It means that the gametes though similar externally yet show functional disparity. They do not have male and female characteristic but positive and negative strains. It is because of this that the gametes of one filament fuse with those from another. The gametes *of Spirogyra* are alike but they are also distinguishable by their degree of motility. The gametes of one filament pass over and combine with the gametes in the other filament. This shows that there must be some difference (not structural) probably biochemical between the two filaments of a conjugating pair. Had there been no such difference there would have been random crossover of cell protoplasts between the two filaments. The active motile gamete may represent a primitive type of male gamete and the passive, non-motile gamete, a primitive type of female gamete. The difference in behaviour between the fusing gametes which are otherwise similar is a step towards anisogamy, a more advanced form of sexual reproduction than isogamy. It is called functional or physiological anisogamy. Anisogamy is more marked when the fusing gametes are of different sizes. They are called the anisogametes and reproduction as anisogamy. The best examples are some species of *Chlamydomonas* (*C braunii*) and *Pandorina.* The anisogametes are produced in vegetative cells called the gametangia. The gametangia may be of different sizes. The smaller ones are called the male gametangia and the larger as female gametangia. The gametes produced by the male and the female gametangia are different in size. The smaller is regarded as male and the larger female. Both are flagellated.

Still more progressive elaboration of the sex organ is found in *Chlamydomonas coccifera* and *Eudorina.* These develop distinct sex organs, the male being called antheridia and the female oogonia. They are specially modified vegetative cells. The gametes produced within them are recognisable as male and female. They are dissimilar in size and shape. The smaller male gamete is active and may be called the sperm. The larger gamete is immobile and may be called the egg. This type of sexual reproduction shows close approach to oogamy. It may be regarded as well marked heterogamy or primitive type of oogamy. In a few species exhibiting heterogamy or primitive type of oogamy the gametes produced in these sex organs are shed (*Fucus*). They fuse externally in water.

The next stage in the evolution of sex is seen in forms like *Oedogonium* which exhibit more advanced type of sexuality. The male gamete is small and flagellated. It is called the sperm. It is produced in the single-celled male sex organ called the antheridium. The female gamete or egg is larger, non-motile and produced singly. It is retained within the oogonium. The antheridia and the oogonia are not specially modified vegetative cells but special sex organs differentiated from the vegetative cells of the thallus. Gametic union no longer takes place in water. The sperms swim to the egg retained within the oogonium to bring about fertilization.

The more advanced condition of oogamy is found in *Vaucheria* and *Chara.* In these forms the elaboration and differentiation of sex organs reaches its culminating point in the green algae. The gametes (sperms and eggs) in these forms are borne in special sex organs (the antheridia and oogonia). These are entirely distinct from the vegetative cells and are not developed from them. They arise as special reproductive branches.

From the above description it is clear that the evolution of sex, which involves the differentiation of gametes, was accompanied by differentiation of sex organs. During differentiation of gametes the male gamete retained motility and small size. The female gamete sacrificed its motility for the increased size and nutritive capacity. This adaptive feature of the female gamete or the egg to store excess food is of advantage to the species. It gives a good start to the organism in life which is of great survival value to the species. Besides, in forms inhabiting moving water isogamy which involves the fusion of similar, motile gametes, must be extremely wasteful. Both the gametes have to expend energy to meet and both being equally active none is able to store much food for the offspring.

The evolution of sexual differentiation in the algae has taken place along several independent lines and not in any one phylogenetic line. One of these is *Volvox* series starting with *Gonium,* passing through *Pandorina, Eudorina* and finally culminating in *Volvox.* In some species of *Volvox,* antheridia and oogonia develop on the same colony and others in different colonies. In the latter case a colony has either antheridia or oogonia but not both. In such species there is distinction into male and female colonies. Sexual reproduction in these forms has evolved to a point where there is sex differentiation.

Ecology of Algae

(i) Aquatic Algae

Majority of algae (about 90 per cent) are aquatic. They may be freshwater (water with low salinity) living either in lentic bodies with stagnant or still water (like lakes, ponds, pools, puddles, tanks, reservoirs, ditches *etc.*) or lotic bodies with running water (like rivers and streams) or marine living in sea water. Some algae may be free floating and free swimming, forming a surface scum on ponds, ditches, *etc.* Along with other forms of plant called phytoplankton, this type furnish food for most aquatic animals, particularly fish. The algae which grow attached to the bottom in shallow water along with the edges of lakes and seas are called benthos (or benthic algae). Some common algae found in still water include: *Oedogonium, Chara, Zygnema, Rivularia.* Those in running water include: *Cladophora* and *Vaucheria,* and those in seas (marine algae) include: *Dictyota, Ceramium, Fucus, Gracilaria, Laminaria,* and Sargassum. Marine algae, most of them being seaweeds, are found in water with high salinity. Many algae (like *Oscillatoria* and *Anabaena*) grow in brackish water that is unpalatable and contains less salt than seawater and more than freshwater. Marine vegetation, which consists largely of algae, occupies only about 2 per cent of the general surface (shoreline and zone of relatively shallow water) of the sea.

(ii) Planktonic Algae

Planktonic algae are the algal forms found floating or weakly swimming in water from one place to the other. They are uniformly distributed in water or may be discontinuous and in the form of patches both horizontally and vertically. These may form a layer of about 1 cm thick and 1-10 m broad. In oceans, the elliptical patches may spread over 60-240 km. In freshwater reservoirs, diatoms are predominant during spring and winter, as blue green algal forms during autumn and summer. In oceans, dinoflagellates are predominant during summer, and diatoms during the rest of the period. The common freshwater planktonic algal forms are: *Chlorella, Scenedesmus, Volvox, Eudorina, Chlamydomonas, Euglena, Microcystis, Anabaena, Aphanothece, Spirulina, Anabaenopsis, Arthrospira, Melosira, Cyclotella, Navicula, Pinnularia, etc.* Among marine planktonic algae the common forms are: *Phalacroma, Dinophyses, Exuviaella, Prorocentrum, Gymnodinium, Peridinium, Gonyaulax, Ceratium, Cyclotella, Sceletonema, Hemidiscus, Fragilaria, Nitzschia, Trichodesmium, Aphanothece, Chlamydomonas, etc.* Usually, marine plankton have a large variety of species than freshwater plankton.

Planktons floating in open water are called *euplanktons,* in rivers *potamoplanktons* and in ponds *heleoplanktons.* Based on their size, plankton have been classified as:

Megaplanktons: Organisms more than 2,000 p, or more in diameter,

Mesoplanktons: Organisms 200-2,000 μm diameter,

Microplanktons: Organisms 20-200 μm diameter,

Nanoplanktons: Organisms 2-20 μm diameter,

Ultrananoplanktons: Organisms less than 2 μ in diameter.

Planktonic species have a relatively slow rate of sinking as they contain gas vacuoles or oil globules which increase their buoyancy. Motility of planktonic colonies is mainly by means of flagella. Wherever phytoplankton are present, they are generally found in very large number, as high as 40 million per litre but this number can be reduced rapidly because they form a regular food of aquatic animals. There is a long list of planktonic algae. They are mainly members of Bacillariophyceae and Dinophyta. Under favourable nutritional conditions the number of some planktonic algae increases enormously forming water blooms.

Water blooms, or phytoplankton blooms, caused by algal blooms, result due to the abundant growth of planktonic algae imparting colour to the water. Algal bloom formation depends on factors, such as temperature increase, availability of inorganic nutrients like nitrogen and phosphorus, and long photoperiod. Algal blooms may be temporary or permanent and pure or mixed. Temporary mixed type, containing several forms, result by simultaneous growth of *Chlamydomonas, Chlorella, Scenedesmus, Pediastrum, Ankistrodesmus etc.* On the other hand, the temporary pure type, containing only one form of algae, often common in inland waters during rains and winter, are produced by *Volvox globator or Chlamydomsonas* spp. Cyanobacteria play a key role in algal bloom formation. Algal blooms containing *Anabaena, Microcystis, Anabaenopsis, etc.*, are a common site in the tropics during late winters and summers. Algal blooms of *Trichodesmium erythraeum,* a cyanobacteria living in sea and having a red pigment phycoerythrine have been a regular phenomenon along the west coast of India. Among permanent type of algal blooms, *Trichodesmium erythraeum* forms permanent red water bloom in Red Sea (due to which probably the sea has been named so) and *Microcystis aeruginosa* blue-green water bloom in temple ponds and other permanent water reservoirs in India. Although these blooms sometimes impart foul odour and make water unpotable, with their accelerated rate of photosynthesis, they form the basis for marine food web and control atmospheric carbon dioxide, a greenhouse gas that in turn can influence heat retention in the earth's atmosphere and cause global warming. However, *Microcystis* and certain dinoflagellates secrete some toxic substances in water.

(iii) Benthic Algae

Benthic algae are aquatic algal forms found attached to one or the other substratum. Among freshwater forms *Cladophora, Pithophora, Chara, Nitella, etc.*, and among marine forms red and brown algae are common. Some benthic algae are annuals (*e.g., Cladophora, Porphyra* and *Enteromorpha*), whereas many of them are perennials (*e.g., Fucus, Sargarssum, Laminaria, Gracilaria, Chondrus,* and *Acetabular.* Among marine benthic forms *Ectocarpus, Ulva, Sphacelaria* and *Acetabularia* are cosmopolitan. The terms given to these algae on the basis of their habitats include *epiphytic* (attached to other plants), *epipelic* (found growing on and in sediments), *epipsammic* (occurring on and in sand) and *epilithic* (growing on rocks). The prolific growth of benthic algae can be witnessed in Gulf of Mannar Biosphere, in southern India and on rocky coasts at Dwarka and Okha on the west coast of India. Large benthic marine algae, the seaweeds are of great economic value.

(iv) Terrestrial Algae (Soil Algae)

Algae found on or in the soil are called terrestrial algae. The soil flora, collectively known as *edaphophytes,* can be spread over *saphophytes* (surface soil flora) and *cryptophytes* or *subterranean* (under soil flora). Some of the soil inhabiting algae can occur in the soil up to the depth of 1 metre or more. These can withstand unfavourable conditions. They thrive under suitable conditions of moisture, illumination and nutritional requirements. Some of the common terrestrial algae are *Oscillatoria, Phormidium, Euglena, Vaucheria, Chlorella* and *Fritschiella.* Some small-size algae live in desert soils as *endedaphic* (in the desert soil), *epidaphic* (on the desert soil surface), *hypolithic* (on the lower surface of stones on desert soil), *chasmolithic* (in rock fissures in desert soil) or endolithic (rock penetrating the desert soil). Soil algae are of great biological and economic importance in as much as they throw light on the evaluation of bryophytes and certain cyanobacteria and are capable of fixing atmospheric nitrogen.

(v) Lithophytic Aigae

These are attached to moist surfaces of stones, rocks or rocky areas, like *Rivularia, Gloeocapsa, Prasiola, Batrachospermum, Enteromorpha, diatoms etc.* Lithophytic algae may be epilithic, living on rock surface (*e.g., Calothrix, Rivularia, Ectocarpus,* and *Polysiphonia*) or endolithic, living inside the rocks (*e.g., Podocapsa,* and *Dalmatella*). *Scytonema* is common on moist walls during rainy season.

(vi) Halophytic Algae, or Eurhaline

These occur in water having high concentration of salts (*e.g., Dunaliella* and *Stephanoptera*) which may sometimes reach as high as 100 parts per thousand. *Chalmydomonas ehrenberghii* can live in water with high salt concentration, where normal plant life is not possible mainly due to exosmosis. Although many occur in saline water or seas, some are found in salt lakes.

(vii) Psammon

Algae occurring on sandy beaches, such as *Vaucheria, Phormidium* and many diatoms, are called psammon.

(viii) Thermal Algae

Algae found in very hot waters (temperature up to 85°C) where normal plant life is not possible are called thermal algae. These are mostly cyanobacteria and occur near hot springs. These algae survive in hot waters mainly because of their primitive morphology, peculiar organization of their protoplasm, absence of a definite nuclear apparatus and mitochondria, and proteins and protein-making mechanism resistant to high temperature (usual protein denatures above 50°C).

Based on the temperature they live in, thermal algae have been classified as *hypothermae* (below 18°C), *hilarothermae* (18-30°C), *euthermae*(30-50°C),*acrothermae* (50-70°C), and *hyperthermae* (above 70°C). As mentioned earlier, majority of thermal algae belong to cyanobacteria (*e.g.*, Synechococcus elongatus, *Mastigocladus laminosus, Phormidium tenue, Conferva thermalis, Hetero hormogonium* sp, *Haplosiphon lignosum, etc.*, and a few to Chlorophyta (Zygnematales) and Bacillariophyceae. These reproduce by means of cell division and fragmentation and rarely by spores.

According to the relict hypothesis proposed by Weed, occurrence of thermal algae in extreme adverse conditions shows the possible existence of such form of life during the archaic period on the earth when it was covered with hot and highly mineralized water. Such plants thus throw light on the origin and evolution of plants on this planet.

(ix) Cryptophytic Algae or Snow Algae

Algae occurring on ice or snow (*Cryoflora*) comprise an important component of cryobiota. Some of these forms impart their own colour to the snow-fed mountains, such as red snow (*Haematococcus nivalis, Ulothrix flaccida* and *Scotiella* sp.), black snow (*Raphidonema*) yellow-green snow (*Chlamydomonas yellowstonensis* in Yellowstone National Park in USA), green snow (*Raphidonema* in European mountains) and purple brown snow (*Ancylonema nordenskioldii*). Some other cryptophytic algae commonly found on permanent snow fields of mountains are the species of *Chlorella* and *Hormidium* and there may also occur *Vaucheria jonesii, Chrysolykos* and *Nannochlories*. Their occurrence under such a condition of low temperature depends on the pH of the ice or snow which is affected mainly by the fine rock dust which penetrates and settles to the ice/snow.

(x) Calcifilic Algae

The algae that are able to bore shells of mollusc, such as Gomontia, are known as calcifilic algae.

(xi) Aerophytic Algae

The algae growing on leaves (epiphyllophytes), bark (epiphoeophytes) and land animals (epizoophytes), are termed as aerophytic algae. Examples are *Phormidium* and *Scytonema.*

(xii) Epiphytic Algae

Algae occurring on other algal members or any other member of the plant kingdom are termed as epiphytic algae. These attach themselves to the other plants only for support and grow there. There is a long list of both freshwater and marine algae growing as epiphytes. A few species of *Oedogonium, Ulothrix, Bulbochaete, etc.*, grow on the other larger freshwater algae. A single marine algal genus *Laminaria* is reported to harbour up to 106 species of epiphytic algae. Some other common examples of epiphytic algae are: needle-shaped diatoms (like *Synedra, Achnanthes* and *Eunotia*) on filamentous algae, such as *Spirogyra, Zygnema, Mougeotia, Oedogonium, Bulbochaete* with mucilagenous surface; and many Cyanophyta (like *Chamaesiphon, Dermocarpa, Lyngbya, Oscillatoria*) on *Cladophora.* Aquatic angiosperms growing along rivers are generally coated with many epiphytic algae, such as diatoms (*Achnanthes, Tabellaria, etc.*) desmids (*Cosmarium*) or Cyanobacteria (like *Nostoc, Oscillatoria* and *Tolypothrix*). *Oedogonium* commonly grows as epiphyte on *Hydrilla* and *Chaetophora* on the leaves of *Vallisnaria* and *Nelumbo.* The algae that grow on leaf surface are called *ephiphyllophytes* (*e.g.*, *Chaetophora* and *Compsopogon*) and those on bark *epiphloephytes* (*e.g.*, *Scytonema, Lyngbya* and *Aphano capsa. The* algae that inhabit tree barks are called *corticolous.*

(xiii) Endophytic Algae (Symbiotic Algae)

Endophytic algae live within or among cells of other plants. *Nostoc* in the thallus of a bryophyte, *Anthoceros; Anabaena cycadearum* in the coralloid root of a gymnosperm, *Cycas; Anabaena azollae* in a pteridophyte, *Azolla* (water fern); and *Nostoc pruniforme* in the stem and petiole of an angiosperm, *Gunnera* are a good example of mutual benefit (symbiosis). *Coleochaetae nitellarum* within the walls of algae *Nitella* and *Endophyton ramosum* in the members of Gigartinaceae also show an endophytic nature.

(xiv) Endozoic and Epizoic Algae

Algae occurring within and on animals are categorized as endozoic or epizoic, respectively. *Stigeoclonium* grows endozoically in the nose of fishes and *Chlorella zoochlorellae* in *Hydra viridis* and sponges. There are about 15 species of Oscillatoriaceae (Cyanobacteria) that occur in the respiratory and digestive tracts of vertebrate animals. Epizoic algae grow on animals like snails, fishes and tortoise. Species *of Stigeoclonium* are found on the gills of fishes and those of *Characiopsis* and *Characium* on the legs of *Branchipus. Cladophora crispata* grows epizoically on the shells *of molluscs.* Some other common endozoic algae are *Euglenomorpha,* which are found in the intestines of the tadpole of *Rana* (frog), *Zooxanthellae* live in intimate association with coral community and *Chlorella*-like algae within *Paramecium* and *molluscs.* Primary importance of inhabiting algae to these organisms is their photosynthetic activity. It was found that C14 labelled products of algae rapidly appeared in the lipids and proteins of animal in association and that sea anemones containing them were photo-tactic while those lacking algae did not move to or from light of varying intensity. An interesting example of association of algae and animals is between a green alga *Platymonas* and flatworm *Convoluta roscoffensis* in which case flatworms are dependent for their development on the alga within it which probably releases amino acids.

(xv) Parasitic Algae

For their food, some algae depend on other plants and animals. These are called parasitic algae. *Cephaleuros,* a green alga, commonly grows on the leaves of angiosperms like *Thea sinensis* (tea), *Magnolia, Rhododendron, Mimusops, Ficus, Hexandra, Psidium, etc.* In the tea plant, *Cephaleuros virescens* causes a disease known as *red rust* disease. Examples of other parasitic algae are a green alga *Phyllosiphon* (on angiosperm *Arisarum vulgar-e*), a green alga *Rhodochytrium* (on the leaves of rageweed), a red alga *Polysiphonia fastigiata* (semiparasite on *Ascophyllum nodosum*), a red alga *Ceratocolax* (on thalli of *Phyllophora*) and a red alga *Callocolax* (on *Callophyllis*). Some Cyanobacteria, such as *Anabaeniolum, Oscillatoria* and *Simonsiella,* occur parasitically in the intestines of some animals. *Chlorochytrium lemnale* occurs as a space parasite inside the tissue of an angiosperm *Lemna. Phyllohium sphagnicolum* occurs as a true parasite on the leaves of a bryophyte *Sphagnum.*

(xvi) Algae in Lichens

Some members of Chlorophyta and Cyanophyta live as *phycobiont* (photobiont) and form a common association with some members of fungi in the form of lichens. Some 22 genera of algae and 12 genera of Cyanobaeteria have been reported to participate in lichen formation. Twenty out of 22 genera, belong to chlorophyta. Two genera, *Trebouxia and Trentepohlia* are the most common phycobionts. Some other examples of algae taking part in the formation of lichens are: *Nostoc, Coccomyxa, Cephaleuros, Urococcus, Phycopeltis, Prasiola, Heterococcus,* and *Scytonema.*

Role of Algae in Soil Fertility

Some of the BGA, were deposited in the soil due to aerobic respiration. They enrich soil fertility. The details of the study discussed below with suitable example.

Blue Green Algae are specialized structures found in the members of Nostocales (except Oscillatoriaceae) and Stigonematales. They arise from somewhat enlarged vegetative cells, and are characterized by the presence of conspicuous pore at one or both poles. In terminal heterocysts the pore is usually present only at one end, whereas intercalary heterocysts have pores at both ends. The position of heterocysts is a specific character; they are terminal in *Gloeotrichia,* intercalary in *Nostoc* and *Scytonema* and rarely lateral as in *Nostochopsis.* They may be solitary or in pairs (*e.g., Anabaenopsis*). Intercalary heterocysts are usually spaced regularly along the filament.

Structure

Heterocysts are thick walled, pale yellow and barrel-shaped structures. The wall is differentiated into an outer fibrous, a middle homogeneous and an inner thin lamellar layer. It is characterised by the presence of pores either at one or both the poles. The wall layers become thicker in the polar region, and the pores are plugged with a refractive material, called polar granules (nodules).

Heterocysts contain dense and homogeneous cytoplasm. They contain mostly carotenoids, hence yellowish in colour. In the absence of phycobilins (which are the principal light absorbing pigments in oxygen-evolving photosynthetic reactions) and chlorophyll a, there is no CO_2 fixation or O_2 production in heterocysts. Thus heterocysts lack oxygen-evolving photosystem II activity.

The photosynthetic lamellae (thylakoids) are tightly packed, with reduced interlamellar spaces. They are concentrated along the periphery in a complex reticulate manner. These lamellae contain two lipids-glycolipids and acyl lipids, which are not found in vegetative cells. Three types of granules (structural granules, crystalline granules, small granules) are associated with the peripheral lamellae, but they decrease in number and eventually disappear as the heterocyst matures. The mature heterocyst may remain attached to the filament for varying periods.

Development

Heterocysts develop from vegetative cells. One or both the daughter cells of a recently divided vegetative cell may behave as proheterocyst A proheterocyst undergoes a number of changes to form heterocyst. The heterocyst is made up of three different wall layers—the outer fibrous and middle homogeneous layers are made up of non-cellulose polysaccharide whereas the inner laminated layer is made up of glycolipids. The outer fibrous layer is uniformly thick, whereas the middle and inner layers are thicker at the polar ends. The pores appear at one or both poles, and through these pores cytoplasmic continuity is maintained with adjacent vegetative cells. There is gradual dissolution of storage granules and breakdown of photosynthetic thylakoids. Sometimes the lamellae may form swollen vesicles and they become concentrated along the periphery and at the pores. In a mature heterocyst the polar pores become plugged with a refractive material (also known as polar nodules) due to which their cytoplasmic connections with adjacent cells are broken.

The above special cell wall layers of the heterocyst permit the atmospheric nitrogen (N_2) to diffuse inside and at the same time stop the atmospheric oxygen to come inside. This is damage control mechanism for the enzyme Nitrogenase which is sensitive to oxygen and low temperature; the enzyme becomes ineffective in the presence of oxygen and at low temperature. The frequency of heterocyst production is Affected by the following external factors:

1. The blue and green light inhibit heterocyst development, whereas the white and red light support it.
2. The concentration of phosphate salts in the medium stimulates heterocyst development, whereas the absence of Mg and Fe ions in the medium inhibits heterocyst development.
3. The presence of combined nitrogen in the medium inhibits heterocyst development. The formation of heterocyst is inversely related to the amount of nitrogen in the medium.

4. The cellular C:N ratio plays an important role in inducing and controlling the differentiation of vegetative cells into heterocysts.

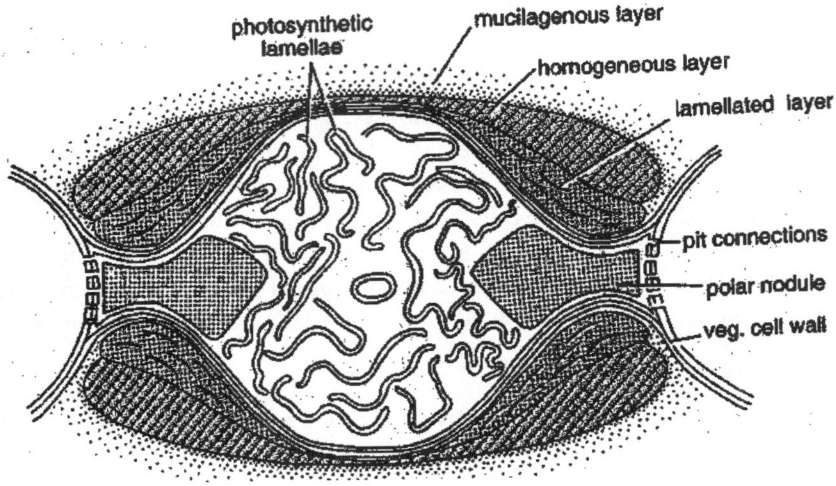

Figure 3.1: Heterocyst: Structure of Mature Heterocyst

Functions

Functions of heterocysts are not definitely known, and several views have been put forward from time to time. Early workers considered heterocysts as simply dead cells, thus having no functions. But this view is no more valid as heterocysts have been shown to respire and germinate just like living cells. Brand (1903) considered them as sporangial or spore like structures and Steinecke (1932) regarded them as secondary reproductive structures. There are evidences to suggest that heterocysts control sporulation. If heterocysts are removed from the trichoma, sporulation is inhibited. Besides this, the vegetative cells adjacent to heterocysts sporulate earlier than those away from it. Canabaeus (1929) found a sort of connection between the formation of gas vacuoles and heterocysts. Fogg (1944) considered them as the sites of salt accumulation. Fritsch (1951) suggested that they secrete certain substances which stimulate growth and cell division.

Fay *et al.* (1968) considered heterocyst as N_2-fixing organ in blue-green algae. This view gets support from the fact that most of the nitrogen fixing blue-green algae possess heterocysts. The factors inhibiting nitrogenase enzyme system also inhibit the formation of heterocyst, and experiments show strong reducing conditions within heterocysts.

Now it has been conclusively shown that heterocysts are the main sites of N_2 fixation in blue-greens, particularly under aerobic conditions. Under anaerobic conditions, the normal vegetative cells have the capability to fix nitrogen, but under aerobic conditions the nitrogenase enzyme becomes inactive due to the presence of oxygen. Under such conditions, the thick-walled heterocysts provide anaerobic environment, necessary for nitrogenase activity. To ensure perfect anaerobic conditions within the heterocysts, even the photosystem II (PSII) is absent from these specialized cells because the PSII does the photolysis of water which generates free oxygen gas. These cells only have PS I which generates assimilatory power (ATP) that helps in nitrogen fixation. Stewart (1970) suggested that the main physiological roles of heterocysts are : (i) control of sporulation, (ii) involvement in growth and nitrogen metabolism, and (iii) archaic reproductive units. The various functions attributed to heterocysts clearly suggest that probably they have varied functions under diverse conditions.

Nitrogen Fixation by Blue-Green Algae

An interesting aspect of blue-green algae is the capacity Of several forms to fix atmospheric nitrogen. This ability of blue-greens makes them truly autotrophic. Beside blue-greens, the ability to fix atmospheric N2 is found in few other prokaryotes like heterotrophic and photosynthetic bacteria.

The ability to fix nitrogen is restricted to about 22 filamentous species of Cyanophyceae. All these forms are characteristically heterocystous, *e.g.*, species of *Anabaena, Aulosira, Nostoc, Tolypothrix, Cylindrospermum, Mastigocladus, Anabaenopsis, Calothrix* and *Scytonema*. Under anaerobic condition, certain non-heterocystous forms (*e.g., Plectonema boryanum*) also fix atmospheric nitrogen.

The blue-green algae, fixing nitrogen from the air, must possess an efficient mechanism to check oxygen inhibition of the enzyme nitrogenase. Thick-walled heterocysts provide a good locus for an efficient anaerobic functioning of nitrogenase, even in aerobic conditions. It has been shown that in heterocystous blue-greens, under aerobic conditions, nitrogen fixation occurs only in heterocysts. But under anaerobic conditions, both heterocysts and vegetative cells possess active nitrogenase enzyme.

The mechanism of nitrogen fixation in blue-green algae has been studied by radioactive tracer technique. As mentioned above, heterocysts, which are nitrogen fixing sites, have nitrogen reducing enzyme nitrogenase. This enzyme reduces nitrogen gas into ammonia, which in turn is metabolized to glutamine by the enzyme glutamine synthetase (GS). Glutamine is then transported from heterocyst to the adjacent vegetative cell, where it is converted into glutamate by the enzyme GOGAT (glutamine oxalo-gluterate amino transferase). Both glutamine and glutamate, by various transamination reactions, form other amino acids. The sequence of reactions in heterocysts and vegetative cells is shown in figure 3. The fixed nitrogen may be utilized in a number of ways : (i) it is assimilated by the blue-greens themselves, (ii) soluble nitrogenous compounds liberate from healthy cells into the culture medium, and (iii) after death and decay, the nitrogenous compounds break into ammonia, which is eventually converted into nitrate by nitrifying bacteria, present in the soil.

In certain blue-green algae (*e.g., Anabaena cylindrica*) the nitrogenase enzyme is also present in vegetative cells, in addition to heterocysts. Besides this, in *Gloeocapsa,* although heterocysts are absent, it has the ability of fixing atmospheric nitrogen. Therefore, it shall be unfair to conclude that only heterocysts are the sites of nitrogen fixation.

In tropical countries, blue-green algae have an important role in increasing the fertility of rice fields. According to Professor R.N. Singh, the fertility of rice fields in India is increased tremendously by species of *Anabaena, Tolypothrix* and *Aulosira. Tolypothrix tenuis* could fix nitrogen at the rate of 406 kg per year per acre. This can increase rice production by about 20 per cent.

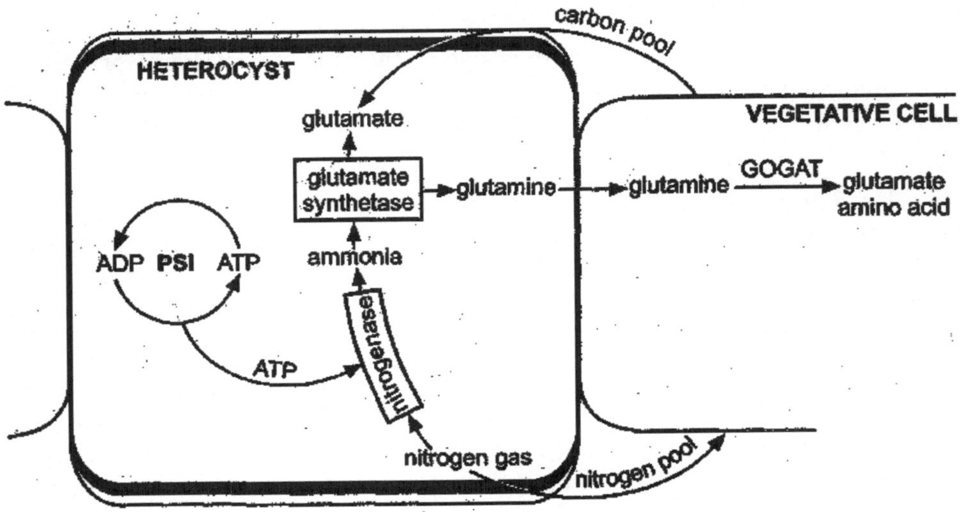

Figure 3.2: Schematic Representation of Nitrogen Transfer from Heterocyst to Vegetative Cell.

Wolk (1996) and Golden and Yoon (1998) have studied the genes involved in early differentiation and pattern of heterocyst formation. They have identified several genes, such as *rtteA, hanA, hetR, hetC, patA,jpatB'*and *patS.* However, little is known concerning control of the cascade of genes whose expression follows induction of differentiation. *NtcA* is required for the nitrogen. fixation under oxygenic growth conditions. *NtcA* hinds to a suitable sequence that is found upstream of the promoter of a number of cyanobacterial genes and is presumed to exert its effect by activating expression. It is required for *hetR* transcription and directly binds to the promoter region of *hetC.* Although it is required for utilization of nitrate and nitrogen, it is not known whether it has a direct role in activation of nitrogenase.

Pat S-5, a product of *patS* gene is synthesized within about 6 hours after the induction of heterocyst in a filament. It represses the differentiation of adjacent vegetative cells of the trichome into heterocysts. This thus helps in maintaining the pattern of spaced heterocysts during growth of the trichome.

There is evidence that heterocysts prevent the differentiation of nearby vegetative cells and also that the Pat S-5 is the inhibitor. However, there is little evidence for direct role of simple nitrogenous compounds (such as nitrate, ammonia, urea or amino acids) in heterocyst differentiation or pattern formation.

Nitrogenase synthesis is tightly regulated by the availability of fixed nitrogen. It, therefore, seems reasonable that fixed nitrogen produced as a result of expression of one nitrogenase would repress subsequent expression of the other.

Culture of Algae

Culture of algae is important in studying the details of external and internal structures, and reproduction of a particular algae or algal group. It is also important to study the details of their life histories, taxonomy, physiology, biochemistry, genetics and molecular studies.

Classification of Algal Culture

The following five types of algal cultures have been categorized by algologists.

(i) **Clonal culture:** The culture obtained by a single cell or fragment of an alga is known as clonal culture. It is used for conducting experiments in genetic homogeneity.

(ii) **Pure (or axenic) culture:** The culture obtained by one strain or an algal species, separated from all other strains, species or other living organisms (*i.e.*, protozoa, fungi and bacteria is called pure or axenic culture.

(iii) **Unialgal culture:** The culture obtained from one strain or species of an alga when other living organisms (protozoa, bacteria and fungi) are also present is called unialgal culture.

(iv) **Enrichment culture:** The culture obtained by adding some nutrients which promote the growth of one or more algal species in a medium, is known as enrichment culture.

(v) **Maintenance culture:** Natural and fresh collection of different algae kept in the laboratory undei proper light conditions is called maintenance culture.

Algae need be cultured in the laboratory for conducting various studies like studies on cytology and genetics, morphological details, physiology and nutrition and ecology of algae. However, it is very difficult to grow pure cultures of algae in the laboratory since the ideal conditions for culturing and maintaining the algal cultures in the laboratory are rather not possible for a long time. Over the years it has been conclusively proved that algal cultures are important in ascertaining the correct identity of the algae besides studying the physiology and biochemistry of algae. Pringsheim (1964) defines cultures as populations which can be maintained indefinitely by subcultures. It was probably Famintzin (1871) who emphasised the utility of inorganic salts for obtaining algal cultures. Later Bouilhac (1897) in addition to inorganic salts used organic compounds also. Initially gelatin was used as solidifying agent but Tischutkin (1897) used agar-agar in place of gelatin. Pringsheim (1912) was the first to obtain pure cultures of several members of Conjugates.

Figure 3.3: Algal Cultures.

Generally algae may be cultured on liquid media, solid media and both solid and liquid media. Some of the media used for algal cultures are Knop's medium, Pringsheim's medium Czurda's medium, Chu 10 medium, modified Chu 10 medium, Bold's basal medium and Ott's artificial sea water medium. These are primary liquid media but can be converted to solid medium by adding agar-agar at the rate of 2 per cent per litre of the medium. Since algae are photoautotrophs, they require the presence of light period alternating with dark periods. However, *Chlorella* requires the presence of continuous light. The pH of the culture medium should be around 7.5 as most algae prefer alkaline pH. Deionised water should be used in the preparation of the culture medium. Cultures should be maintained at a temperature between 20°C and 30°C. Generally bacteria contaminate the algal cultures. To obtain bacteria less cultures, antibiotics like aureomycin is used. Before inoculation treatment of algal suspensions with 0.01 per cent Mercuric chloride has been found useful.

Almost all the algae including marine algae have been grown in pure cultures. Blue green algae are the easiest to be grown followed by green algae.

The Potential for the Cultivation of Seaweeds

Seaweed cultivation by Rope Methods and Tank culture for maniculture programe to get higher yield of seaweed for production of Agar-Agar.

The Indian coastline stretches to 8129 km, excluding the creekareas with tidal effect another 2500 km. The diversity of macroalgae or seaweeds approaches 900 and compares well with that of the Indian Ocean region—Malay Peninsula to Philippines. The productivity in the region at the fertile sites measured around 30 tons, $ha^{-1}.yr^{-1}$. On the Indian coast the major productive areas are located in Tamil Nadu, Gujarat, coral islands of Andaman and Nocobar, and Lakshadweep. As per the recent assessment, the standing stock of seaweeds measured 4.33 lakh tons (wet). Of these agarophytes (*Gelidiella acerosa* and *Gracilaria* spp.) carrageenophytes (*Hypneaspp.*) constitute 2 per cent each, alginophytes (*Sargassum* and *Turbinaria* spp. 5 per cent, and others 91 per cent (Subbaramaiah *et al.*, 2006, Umamaheswara Rao, 2009). The Tamil Nadu resources accounted for 101381 tons (wet). Utilization of the resources has so far been mainly for the production of agar-agar, algin and fertilizers. The red algae *Gelidiella acerosa, Gracilaria edulis, G.salicornia, G.foliifera* and *G.verrucosa* are used for agar production, and the brown algae *Sargassum wighti S.ilicifolium, S.vulgare, Sargassum* spp. *Turbinaria* and *Cystoseira trinodis* for algin and liquid fertilizer. The present requirement of 400 tons agar and 1000 tons algin annually, only 1/3 is met by Indian industries. About 800 tons dry agarophytes and 2000 tons dry alginophytes drawn from the natural beds now reach the industrial units.

The seashore areas in the Palk Bay (Athankarai to Rameswaram, and the Gulf of Manner (GOM) region (Dhanushkodi to Rameswaram), as the Coral islands endowed with lagoons and coral reefs in the GOM present ideal conditions for the growth and natural production of seaweed beds. The shallow sea of the inshore waters also supports the growth of seagrass meadows. The topography and seawater environment with minimum tidal variation greatly favour macroalgal production. Major aspects of the biology propagation in the natural beds and agar Recovery of seaweed beds in the intertidal and near shore areas appears remote unless a complete ban on seaweed harvesting is enforced atieast for one year. Subbaramaiah (1977) addressed these problems.

Seaweed suppliers have organized bulk collection of the economic seaweeds. Collecting, drying, storage of the raw seaweed material and supply to the industrial units have become a lucrative business providing livelihood to several thousand fisherfolk. The seaweed beds native to the shores from Rameswaram to Kanyakumari and from the many islands in the GOM were subjected to successive harvests year after year. The harvest data was gathered by the CMFRI researchers from the Seaweed Landing Centres. The agarophytes exploited during the years from 1979 to 2004 varried from 248 to 1518 tons (dry). The extent of seaweed collections depended upon their availability on the one land to the requirements of the industrial units on the other. The time and period of seaweed collection also largely depended on the Production Plant units. The seaweed exploitation attained a peak period during 1998 to 2002. Later years marked a decrease in seaweed supplies resulting in the closure of some agar units (Kaliaperumal, 2006).

Thivy, Umamaheswara Rao and others made pioneering algal studies in the CMFRI, Mandapam right from 1960 onwards. A cottage industry process for agar production using *Gelidiella* as raw material was worked out by Thivy (1960). Initially, commercial agar production started as a cottage industry. Later CSMCRI technology for agar production (Kappanna and Rao, 1963) was adopted by the industry. Industrial agar units increased in number during 1970-75, Madurai alone contributing to the maximum (say a dozen). At present in all 10 agar and 12 alginate factories are located in Tamil Nadu, Kerala and Karnataka states.

Algal Pigments and their Uses

1. Chlorophyll

There are five types of chlorophylls, *viz*. chl. a, b, c, d and e. Of these, chlorophyll-a is present in all groups of algae, chlorophyll **b** only in Chlorophyceae and Euglenophyceae, chlorophyll **c** largely in algae of marine habitats (Phaeophyceae, Cryptophyceae, Bacillariophyceae and Chrysophyceae), chlorophyll **d** in some red algae only as a trace constituent and chlorophyll **e** in certain Xanthophyceae such as *Vaucheria hamata* and *Tribonema bombycinum.*

2. Xanthophyll

More than 20 types of xanthophylls are formed by the incorporation of molecular oxygen into carotene molecule. Many xanthophylls, common in higher plants (lutin, violoxanthin and neoxanthin) are found in the members of the Chlorophyceae and Phaeophyceae, Fucoxanthin is the main xanthophylls pigment of Phaeophyceae and diatoms whereas myxoxanthophyll, myxoxanthin and ocilloxanthin are found only *in* the Cyanophyceae.

3. Carotenes

These are oxygen free alicyclic compounds of isopyrene units. The five types of carotenes occur in algae are a - carotene in Chlorophyceae, Cryptophyceae and Rhodophyceae, b - carotene in all algal groups, except Cryptophyeae, c - carotene in Chlorophyceae, d - carotene in Bacillariophyceae, Cryptophyceae, Phaeophyceae and Cynophyceae and flavacene in members of Cyanophyceae,

4. Phycobilins

These are water-soluble complexes of protein and bile pigments present in photosynthetic tissue of plants. Phycobilins are red (Phycoerythrin) and blue (Phycocyanin) pigments, which are confined to Rhodophyceae and Cyanophyceae respectively. They act as light harvesting pigments in photosynthesis and the light absorbed by them to transfer to chlorophyll a. Thus like carotenoids, phycobilins are also accessory pigments.

Of the external features, colour of the algal thallus is most impressive and distinctive. It may be green, yellow, red, brown, and other shades of these. These striking differences of colour in thalli of Algae form one of the basic criteria for a preliminary classification. The colour of the thallus is due to the presence of definite chemical compounds called the pigments. Each pigment has its own characteristic colour. The particular colour that a thallus has may be due to the predominance of one in a combination of several others. Each algal division has its own particular combination of pigments and a characteristic colour. In all, there are four different kinds of pigments found in the Algae. These are five chlorophylls, 20 xanthophylls, five carotenes and phycobilins. Usually the algal pigments are located in the plastids. The latter may be ribbon, band or cup-like in form or a parietal plate. In some they are lens-shaped, disc or network like or an axial band, a stellate or an oval body, a lobed disc or a parietal ring. The Cyanophyta lack plastids, the pigments are located in the lamellae.

The chlorophylls. There are five known chlorophylls, namely, chlorophyll *-a, -b, -c, -d* and *-e.* They show different spectral properties. Of these chlorophyll-*a* occurs in all Algae. Chlorophyll-*b* is found in Chlorophyta, Euglenophyta and Charophyta. Chlorophyll-*c* occurs in Bacillariophyta, Pyrrophyta and Phaeophyta. Chlorophyll-*d* occurs only in red algae. Chlorophyll-e has been found in the Xanthophyta (*Tribonema,* and zoospores of *Vaucheria*). The green colour of the thallus in green algae is due to the presence of two closely related pigments chlorophyll-*a* and chlorophyll-*b*. They are the dominant pigments in Chlorophyta, Charophyta and Euglenophyta. Commonly we call these two pigments simply as chlorophyll. In the cell the pigments are contained in a specialised part or parts of the cell protoplast called the plastids. The plastids containing both chlorophyll-a and *b* are called the chloroplasts and those which lack chlorophyll *b* and have carotenoids in excess over chlorophyll are usually called chromatophores. In the green algae *chlorophyll-a* and *chlorophyll-b* make up about two-thirds of the total pigment in the chloroplast. Associated with these are the yellow carotenes and yellow to orange brown xanthophylls. There are two carotene pigments namely *a*-carotene and carotene and eight or nine xanthophylls. The chlorophylls, carotenes and xanthophylls are present in different proportions. This accounts for the various external colours of thalli in the Chlorophyta.

Chlorophyll-a and *chlorophyll-b* are microcrystalline solids. The empirical formula for chlorophyll-*a* is $C_{55}H_{72}O_5N_4Mg$ and for chlorophyll-*b* is $C_{55}H_{70}O_6N_4Mg$. *Chlorophyll-a* and *-b* are not soluble in water even under prolonged boiling. They are fat soluble and dissolve in either, benzol, acetone, chloroform, carbon bisulphide and ethyl alcohol. *Chlorophyll-a* gives a bluish green coloration if dissolved in the above solvents whereas *Chlorophyll-b* gives a pure yellowish green coloration. The Chloropohylls absorb blue and red rays and are the import photosynthetic pigments.

Carotencids are powerful antioxidants. Recent studies have shown the correlation between a diet rich in carotenoids and a diminishing risk of cardio-vascular disease,cancer (p - carotene, lycopene), as well as opthamological diseases (lutein, zeaxanthin). Brown seaweeds are particularly rich in carotenoids especially in fucoxanthin, p-carotene, and violaxanthin. The main carotenoids in the red algae are the P-carotene and do-carotene and their dihydroxlated derivatives. A lot of studies demonstrated the antioxidant properties of red seaweed carotenoids and the role they play in preventing many pathologies linked to oxidative stress. An interesting group of high-value macro algal products is the phyco-biliproteins. These protem-containing pigments, which are unique to certain algae, serve as valuable fluorescent tags with many applications in high-technology areas, such as flow cytometry, fluorescence-activated cell sorting, and histochemistry. The major product in this area is r-phycoerythrin, which is currently derived from species of *Porphyra,* either cultured or harvested from the wild. Because R-phycoerythrin exists as part of a mixture of several different phyco-biliproteins, sophisticated separation and purification procedures, usually involving chromatography, are required for the production of the pure product.

The chlorophyll pigments arc formed from a substance called protochlorophyll which is synthesised in the dark, it is generally changed to chlorophyll only in the light The carotenoids. It is a group of yellow, orange, red, and brown pigments. About sixty different carotenoids have been reported in plants. They are placed under two categories, the orange yellow carotenes and yellow or brown xanthophylls or carotenols. The carotenoids are protective pigments functioning as screens to light. In addition they pass on energy of light waves to chlorophyll. They absorb blue and green light waves. There is a suggestion that the carotenoids function as co-enzymes in the photosynihelic process.

(a) Carotenes

The carotenes are linear unsaturated hydrocarbons represented by a chemical formula $C_{40}H_{56}$. There are five carotenes so far known, namely, carotene *-a, -B, -e, -y* find lycopene. Of these *B*-carotene is present in most algae. It is replaced by cartone-*a* in the Siphonales (Caulerpales). Cryptophyta and to a lesser extent in red algae. The Charophyta also lack *B*-carotene. It is replaced by *e*-carotene and lycopene. The Bacillariophyta possess both *B*- and c-carotene. Carotenes arc fat-soluble pigments. They are insoluble in aqueous solution but are soluble in lipid solvents such as ethyl alcohol, chloroform and carbon bisulphide acid absorb blue and green light waves.

The resting cells of some of the Algae (terrestrial species of *Trentepohilia*) contain carotenoid pigments outside the chloroplasts. The eyes-pols of zoospores and other motile stages, also contain carotenoid pigments. The haematochrome, or the yellow and the red coloured substances are according to some scientists (Zopf and Wisselingh) represented by two or more carotenoid pigments. The haematochrome in *Sphaerelia* appears only when the surrounding medium lacks nitrogen (Pringsheim), The same fact has been observed in some of the Chlorophyceae growing in boggy places or the moorlands. Such algae are yellow in colour due to the fact that the moorlands usually lack nitrogen contents.

(b) Xanthophylls

They are yellow or brown pigment represented by a molecular formula $C_{10}H_{56}O_2$. They are closely related to the carotenes but contain oxygen in addition lo carbon and hydrogen. Both carotenes and xanthophylls are not watersoluble but are soluble in chloroform. There are many xanthophylls so far known in algae. The chief among them are *lutein, Siphonein, Zeuxanthin. Siphonoxanthin, Astaxanthin, Neoxonthin, Crypioxanthin, Vioiaxanthin, Lycopene, Diaioxanthin. Diadinoxanthin, Flavoxcmihin. Myxoxanthin. Dinaxanthin, I'araxanfhin, Antheraxanthin. Osrilloxanthin, Myxoxanthophyll, Flavacin, Peridinin, Neo/ucoxartthin-A arid -B and Fucoxanthin,* Of these Fucoxanthin is a characteristic pigment of the Phaeophyta imparting distinctive brown or olive brown coloration to the thallus. The diatoms too contain several brownish xanthophylls such as fucoxanthin, diatoxanthin and a few others. Peridinin is found only in the Pyriophyta. Myxoxanthin and a few others are reported only in the Cyanophyta. Taraxanthin is characteristic of red algae and antheraxanthin of Euglenophyta. like carotenes the xanthophylls absorb blue and green light rays and are fat soluble. Fucoxanthin is considered a supplementary light absorbing pigment in Phaeophyta and Bacillariophyeceae. An impure solution of fucoxanthin can be extracted from the dead specimens in water and a chemical formula, $C_{40}H_{50}O_6$ has been given to it by Willstatter and Page. *Ectocarpus, Dictyota* and *Laminaria* contain a good amount of fucoxanthin in their chromatophores. In *Fucus* the amount of this pigment is comparatively less.

The Phycobilins. It is another group of pigments comprising the tetrapyrrolic compounds joined to globulin-proteins. So far seven phycobilin pigments both blue and red have been enlisted. They are *Phycoerythrin -r. -c, -x, -b* and *Phycocyanin -r* and *-c.* The seventh is *Allophycocyanin.* The phycobilins are water-soluble pigments found in the red and blue green algae. Of the seven phycobilins. r-phycoerythrin and r-phycocyanin are common. The former absorbs blue, green and sometimes yellow rays whereas the latter absorbs green, and sometimes yellow rays whereas the latter absorbs green, red and yellow. Both the phycoerythrin and phycocyanin are closely related chemically. They are colloidal proteinaceous pigments which can be extracted from the cell in hot water. During extraction procedure of phyeobilins the two components (pigment and protein moiety) cannot be separated. To indicate the existence of the pigment protein complex the name of the phycobilin pigments was changed to biliproteins. Both the phycoerythrin and phycocyanin pigments are found in the red algae and blue-green algae. Those of the former are of r-type and those of blue-green algae c-type. Chlorophyll content in them is quite low. In some forms small quantities of chlorophyll-*a* has been reported. Carotene and xanthophyll pigments are also present. Red algae occurring in colder regions show higher content of chlorophyll-a. Phycoerythrin crystallises in hexagonal prisms whereas crystals of phycocyanin are rhombic plates. The blue pigment allophycocyanin is present in the blue-green algae only. It is absent in the red algae. The deep red colour exhibited by some of the deep water Florideae is due to the higher concentration of phycoerythrin in the chromatophores. Phycocyanin is either lacking or present in negligible amounts. Some fresh water Florideae like *Batrachospermum* contain a large quantity of phycocyanin and small quantity of phycoerythrin. Their colour is blue-green. Red coloration is evident only after death. *Batrachospermum vagum* is believed to lack phycoerythrin. Phycobilins differ from chlorophylls and carotenoids in being water soluble.

Chlorophyll-*a* is of prime importance in photosynthesis. The accessory pigments function only indirectly. The wavelengths of light which are not absorbed by chlorophyll are absorbed by Phycocyanin and Phycoerythrin. The light energy trapped by the latter two pigments is then transferred to chlorophyll-*a* which utilizes it in food synthesis.

Water Bloom

Water blooms or Blooms are formed by the growth of a solitary alga, rarely by a few algae. When the algae are abundant in a pond or lake as to give it a distinct colour to the entire body, such an algal growth is called as **water blooms.** The colour of the algae determines the colour of the water blooms. The algae involved in the development of water blooms are generally planktons.

Water blooms are generally formed by the rapid growth and multiplication of algae which usually belong to class Cyanophyceae. A few members of Chlorophyceae, Chrysophyceae, Euglenophyceae and Pyrophyceae are also known to form water blooms. Water blooms are formed in fresh waters, inland salt lakes and in the sea.

Water blooms may be temporary or permanent. Generally a number of temple ponds and a few lakes and big ponds have permanent water blooms. The water blooms are usually thick during winters and summers while during the rainy season the blooms become thin or get diluted because of sudden splashes of rain drops that dilute the blooms.

Figure 3.5: Algal Blooms.

In India, the water blooms in temple ponds, lakes *etc.* are formed of *Microcystic aeruginosa* and *M.flos aquae.* Water bloom forming algae of class Cyanophyceae include species of *Anabaenopsis, Spirulina, Oscillatoria, Arthrospira etc.*, those of Chlorophyceae are *Chlamydomonas, Pandorina, Euderina, Gonium, Volvox, Cosmarium, Closterium etc. Euglena* of Euglenophyceae, *Synura* of Chrysophyceae,

Peridinium, Ceratium and *Gymnodinium* of Pyrrophyceae. In salt takes of Sambhar, where pH ranges between 14 and 17, blooms are formed *of Anabaenopsis.* In the esturine regions of rivers in South India Colonies of *Anaebaenopsis arnoldi* v. *indica* form the blooms. In the India seas, *Trichodesmium erythraeum* and *T. thiebantic* are the bloom forming algae.

Although bloom forming factors are yet to be understood, it is generally believed that optimum conditions necessary for sound algal growth must be present in water bodies. Water blooms formed of Cyanophycean algae are good nitrogen fixers and therefore can be used as fertilisers. Water blooms also provide food for immense flocks of flamingos and other migratory birds. However, water blooms may be harmful because they are indirectly responsible for fish mortality due to depletion of oxygen. Sometimes they are responsible for unpleasant taste to drinking water and foul smell. These also liberate deleterious substances to aquatic animals. Water blooms can be controlled by application of copper sulphate or Chlorine. Cyanophages have also been used to control blooms besides algicides.

Economic Importance of Algae

1. Algae as Food

Marine algae as food: The earliest records are those of the Chinese, who mentioned such food plants as *Gracilaria* and *Porphyra* in their **"Materia medica"** several thousand years ago. *Porphyra* is a widespread plant in the North pacific. It grows in the northest from the Gulf of Alska to California and is reported by anthropologists to have been used for thousands of years by the American Indians ever since they came from Siberia. Because of the Indians aversion to salting their food, they used *Porphyra* to supply the salt needed for their bodies. Perhaps the best known and most widely used food alga in Western Europe in recent centuries, was Irish moss or carrageen *[Chondrus crispus*) which was cooked with milk, seasoned with vanilla or fruit, and made into a highly plate be dish known as balancemanges.

Several algal forms are used as food due to high nutritive value. The common examples are *Porphyra, alaria, Viva, Chlorella, Chondnis, Rhodymenia* and *Codiam etc. Porphyra* has very rich in protein (30-35 per cent) and carbohydrates (40-45 per cent) and also a good source *of* vitamin B and C. It is a common food item in Japan alone more than 30 million kg of *Porphyra* is used per year as food.

A gelatinous carbohydrate obtained from *Chondnis crispus* is used in puddings. It is also used as a stabilizer and clearing agent in the production of beer. A salty confection commonly known as dulse is obtained from *Rhodymenia palmetto.* In Scotland it is chevied as tobacco. *Lemanea* is considered a delicacy in the Manipur region of India. Food obtained from *Laminaria saccharina* is rich in carbohydrates (57 per cent). Some red algae such as *Gelidium,* and *Gracilaria* are used in ice cream and jelly.

Chlorella, nutritional value is comparable to that of a mixture of soybeans and spinach. It contains about 30 per cent carbohydrates, 30 per cent proteins and 15 per cent lipids. And it is highly is useful in space flights and nuclear submarines as food/and so also in recycling and oxygen regenerating system. In India, at CFTRI, Mysore, research is being conducted on the use of blue green alga, *Spirulina* as a supplement to diet. It contains 60 per cent protein, essential vitamins and unsaturated fatty acids. *Scenedesmus,* besides being rich in protein^. is also a rich source of threonine. It has a high dieted value equal to that of "skim milk. The colonies of *Nostoc* commune are boiled and are used as food in China and Java.

2. Polysaccharides

Of all seaweed products, hydrocolloids have the biggest influence on modern western societies. They have attained commercial significance through their use in various industries which exploit their physical properties such as gelling, water-retention and their ability to emulsify. Seaweeds contain large amounts of polysaccharides, notably cell wall structural, but also mucopolysaccharides and storage polysaccharides. Polysaccharides are polymers of simple sugars (monosaccharides) linked together by glycosidic bonds, and they have numerous commercial applications in products such as stabilisers, thickeners, emulsifiers, food, feed, beverages, *etc.* Green algae contain sulphuric acid polysaccharides, sulphated galactans and xylans, brown algae alginic acid, fucoidan (sulphated fucose), laminarin (p-1,3 glucan) and sargassan and red algae agars, carrageenans, xylans, floridean starch (amylopectin-like glucan), water-soluble sulphated galactan, as well as porphyran as mucopolysaccharides located in the intercellular spaces. The three major groups of phycocolloids are alginates, carrageenan and agar, and these are used for applications such as thickening aqueous solutions, forming gels and water-soluble films, and stabilizing products such as ice cream, tooth paste, mayonnaise, *etc.*

3. Algae as Fodder

In several European countries algae such as *Fucus, Lamindria, Ascophyllum* are being possessed into cattle feed. Besides plant like *Sargassum* and *Macrocystis are* being used as a fodder in countries like China.

Seaweeds are utilized as animal feed in some countries. They are cheap sources of minerals and trace elements. Hence, the meal prepared from seaweeds can be given as supplements to the daily rations of cattle, poultry and other farm animals. Seaweed meal can be obtained by grinding cleaned and washed seaweeds such as *Ulva, EnteromorpJia, Sargassum,Padina, Dictyota, Gracilaria* and *Hypnea.* Seaweed meal can also be mixed with fish meal and used as poultry feed.

4. Algae in Industry

Many products *of* commercial and pharmaceutical importance have been derived from algae. Agar, carrageenin, alginate and diatomite are sline important commercial products of algal origin.

5. Agar-Agar

It consists of two polysaccharides, agarose and agaropectin. Agar is obtained commercially from species of *Gelidium, Pterccladia, Gracilaria, Acanthopeltis, Ahnfeltia, Chondrus* and *Gigartina.* The greatest use of agar is in food, pharmaceutical and cosmetic industry. It is used, in the manufacture of processed cheese, puddings, creams and jellies. In preservation and canning of fish and meat it is used as gelling and. thickening agent and to reduce the undesirable effects. Pharmaceutically, agar is used as a laxative. It is an important constituent of cosmetics (ointments, lotions, *etc.*). It is used for almost a century as stiffening agent in culture media.

Agar is the major constituent of the cell walls of certain red algae, especially members of the Gelidiaceae and Gracilariaceae. Agar is a mixture of polysaccharides, which can be composed of agarose and agropectin, with similar structural and functional properties as carrageenans. Agar is soluble in water and extracted in hot water. It gels at room temperature and freezing and thawing are used to purify. The important and commonly occurring agarophytes of India are *Geliediella acerosa, Gracilaria edulis, G. crassa, G. corticata,G. foliifera* and *Acanthaphora* sps. Agar-agar is used in the manufacture of capsules for medical applications and as a medium for cell cultures, *etc.*

6. Carrageenin

It is a phycocolioid (polysaccharide colloid formed by an alga) similar to agar. Carrageenin is obtained from the cell walls of *Chondrus crispus* (Irish moss) and *Gigartina* cultured as a source of carrageenin. It is extensively used in stabilization of emulsions in paints and cosmetics. In alcohol and sugar industry it is used as a clearing agent. In recent years carrageenin has supplemented agar in food and pharmaceutical industry. It is also utilized in the textile, leather and brewing industries.

Carrageenans are a group of biomolecules composed of linear polysaccharide chains with sulphate half-esters attached to the sugar unit. Their unifying characteristic is that they are all made up of galactose-related monomers (a-l,3-D-galactose and P-1,4-3,6-anhydro-D-galactose) to which sulfate groups are attached. There are three general forms (kappa, lambda and iota), each with its own gelling property. They are primarily derived from *Kappaphycus alvarezii., Chondrus crispus,* and *Eucheuma spinosum.* Like the alginates, the carrageenans can be used to gel, thicken, suspend, and stabilize foods and other products. Food applications for carrageenans are many, including canned foods, dessert mousses, salad dressings, bakery fillings, ice cream, instant desserts and canned pet foods. Industrial applications for purified extracts of carrageenans are equally diverse. They are used in the brewing industry for clarifying beer, wines and honeys, although less commonly than previously. From a human health perspective, it has been reported that carrageenans have anti-tumour and antiviral properties. Carrageenans are also used as suspension agents and stabilisers in other drugs, lotions and medicinal creams. Other medical applications are as an anticoagulant in blood products and for the treatment of bowel problems such as diarrhoea, constipation and dysentery.

They are also used to make internal poultices to control stomach ulcers. The use of carrageenan for food applications started almost 600 years ago. The WHO Joint Expert Committee of Food Additives has concluded that it is not necessary to specify an acceptable daily intake limit for carrageenans. Furthermore, carrageenans can cause reproducible inflamatory reaction and remain a standard irritant for exarnining acute inflammation and anti-inflamatory drugs.

7. Alginate

These are salts of alginic acid, which occur in the cell wall (particularly middle lamella and primary, wall) of the brown algae belonging to the order Laminariales. Alginate is extracted from *Laminaria, Ascophyllum, Fucus, Macrocystis, Ecklonia, Lessonia* and *durvillea.* Alginates are non-toxic and viscous and readily form gel, hence useful as thickener (in ice-cream, synthetic cream, polish, emulsion paints) and gelling agent (in confectionery and dental impression powder). Alginic acid used for stops bleeding effectively. Flame proof fabrics are also prepared from alginates. When mixed with resin, alginates are used for binding panel boards.

Algins/alginates are extracted from brown seaweed and are available in both acid and salt forms. The acid form is a linear polyuronic acid and is referred to as a alginic acid, whereas the salf form is an important cell wall component in all brown seaweed, constituting up to 40-47 per cent of the dry weight of algal biomass. *Sargassum, Turbinaria, Dictyota, Padina, Cystosiera, Hormophysa, Colpomenia, Spatoglosum* and *Stoecospermum* are the align yielding seaweeds occurring in the Indian waters, Algin is extensively used in industries related to food processing, pharmaceuticals, feed and cosmetics, rubber products, textile products, adhesives, paper products and other miscellaneous products. It has been reported that alginic acid helps to protect against potential carcinogens and they clear the digestive system and protect the surface membranes of the stomach and intestine. They absorb substances such as cholesterol, which are then eliminated from the digestive system and result in hypocholesterolemic and hypolipidemic responses. Furthermore, the algins have anticancer properties and antitoxic effects on hapetitis.

8. Fucoidan

Fucoidan/fucans/fucanoids are a group of polysaccharides (fucans) primarily composed of sulphated L-fucose with <10 per cent of other monosaccharides. They are widely found in the cell walls of brown seaweed, but not in other algae or higher plants and have several biological activities. Their use as drugs would offer the advantage of no potential risk of contamination with viruses or particles such as prions.

9. Mannitol

Mannitol is an important sugar alcohol of the hexite series found in brown algae. Mannitol is a constituent of cell sap. It occurs also as mannitan. The chief raw materials for the extraction of mannitol are *Fucus vesiculosus, Laminaria hyperborea,*

Ecklonia radiata, Bifurcaria brassiformis, Sargassum spp, and *Turbinaria* spp. In the extraction of mannitol, thedried brown seaweed materials are pre-treated with dilute HCI (10-15 per cent). The aqueous acid extract after neutralizationis evaporated to dryness. From this mixture of salts, mannitol and soluble polysaccharides are extracted with boiling methanol for 5 hours in the Hannen and Badum extractor. The solution containing the extracted material is allowed to stand for 24 hours at 5°C and the precipitate a crystalline mannitol is filtered and dried before weighing. In pharmacy, mannitol is applied for the production of tablets. Mannitol is also used for making diabetic food, chewing gum *etc.* Mannitol is employed as dusting powder in the paint and varnish industry, leather and paper industry, pyro-techniques and in making explosives. In organic synthesis and in plastic production, mannitol is used as plasticizers for the production of resins.

10. Nutritional Value

The uses of seaweed as food have been traced back to the fourth century in Japan and the sixth century in China. Today those two countries and the Republic of Korea are the largest consumers of seaweeds as food. Seaweeds contain nutritional elements such as proteins, lipids, carbohydrates, vitamins and minerals, the content of these elements varies depending on season and the area of production. The ash content in seaweed is high compared to vegetables. Seaweeds are low in calories from "a nutritional perspective. The lipid content is low, and even though the carbohydrate content is high, most of this is dietary fibres and not taken up by the human body. However, dietary fibres are good for human health. Fresh, dried and processed seaweeds are utilized for human consumption. They are eaten as salad, curry, soup or jam.

11. Algae as Antibiotics and in Medicines

Many algae such as *Chlorella, Cladophora, Lyngbya, Polysiphonia, Laminaria* and *Halidrys* synthesize antibiotic substances. The first such antibiotic, **chlorellin,** was obtained from *Chlorella.* Extract of *Ascophyllum nodosum* is effective against both Gram-positive and Gram-negative bacteria. Water passed through a filter containing *Nitzschia palea* showed antibiotic property. The nitrate is effective only if the algal filter is in light and photosynthesising. It suggests that antibiotic-substances are formed during photosynthesis.

Members of the order Charales have larvicidal property hence may be useful in controlling mosquitoes. Extracts from *Alsidium, Corallina, Codium, Durvillea, etc.* are effective vermifuge.

Tse-ko-Tsoi, and anthelmintic drug, is prepared from a red alga, *Digenia* simplex, in South China. Some algae are used in the treatment of diseases of kidney, urinary bladder and lungs. Brown algae are used in various goiter medicines because of their high iodine contents. *Gelidium* is useful in stomach disorders.

Seaweeds are a good source of a number of vitamins. The diatom *Nitzschia* is fairly rich in vitamin A, riboflavin is present in good amount in *porphyra tenera,*

Gelidium amansiiand Chondrus spp., Furcellania fastigata and *Rhodomela subfusca* are rich in thiamine, and vitamin C content of *Porphyra lacinata* is higher than that of oranges.

Seaweeds were considered to be of medicinal value in the orient as early as 3000 B.C. The Chinese and Japanese used them in the treatment of goiter and other diseases. The Romans used the seaweeds for healing wounds, burns and rashes. The British used *Porphyra* to prevent scurvy (vitamin-C deficiency disease) during long voyages. Various red algae particularly *Corallina officinalis* and *Alsidium lielminthocorton* are employed as vermifuges. Some other red algae such as *Chondrus, Gracilaria, Gelidium* have been used to treat various stomach and intestinal disorders. *Laminaria* is used as a pain killer and also to distend the uterus. Carrageenan may be useful in ulcer therapy and the alginates are found to prolong the rate of activity in certain drugs. Species of *Sargassum* are used for cooling and blood cleaning agent. *Hynea musciformis* is used as a vermifuge or worm expelling agent. The iodine rich seaweed *Sarconema* can be used for controlling the goiter, disease caused by the enlargement of thyroid gland. However, in recent years pharmaceutical firms have started looking towards marine organisms, including seaweeds, in their search for new drugs from natural products. Products from seaweeds increasingly being used in medical and biochemical research.

Gelidium very early became employed for stomach disorders and tor heat -induced illness. Perhaps the algae used most widely and for the longest time for medicinal purposes have been the agarphytes, including *Gelidium, Pterocladia* and *Gracliaria.* The name "agar agar" is of Malay origin and means "jelly". Boiling up seaweed and cooling the resulting liquid obtained this jelly. Agar early became useful for stomach disorders and as a laxative, and was once employed as a slenderising dietetic. Agar is used in the cultivation of bacteria. Irish moss {*Chondrus crispus*) has had a long medical history in Europe in the treatment of diarrhea, urinary disorders, and chronic pectoral infections. It is a good remeady for tuberculosis in lungs, *Corallina officinalis,* a small red algae popularly known as vermifuge algae, since it kills small insects or worms inside the stomach of human beings.

12. Protein and Amino Acid

Protein content of brown seaweeds is generally small (average: 5-15 per cent of the dry weight), whereas higher protein contents are recorded in green and red seaweeds (on average: 10-30 per cent of the dry weight). In some red seaweed, such as *Palmaria-palmata*(dulse) and *Porphyra tenera* (nori), proteins can represent up to 35 and 47 per cent of the dry matter, respectively. Most seaweed species contain all the essential amino acids and are a rich source of the acidic amino acids, aspartic acid and glutamic acid

13. Lipid and Fatty Acids

Lipids represents only 1-5 per cent of algal dry matter and show an interesting polyunsaturated fatty acid composition particularly regarding with omega 3 and omega 6 acids which play an important role in the cardio vascular diseases, osteoarthritis and diabetes. The green algae show interesting levels of alpha linoleic acid ($\dot{\omega}3$ C 18:3). The red and brown algae are particularly rich in fatty acids with 20 carbon atoms: eicosa pentanoic acid (EPA, $\dot{\omega}3$ C20:5) and arachiodonic acid (AA, $\dot{\omega}6$ C20:4).

14. Polyphenols

Seaweed polyphenols also called phlorotannins differ from terrestrial plant derived for phloroglucino units (1,3,5-trihydroxybenzine). Phlorotannins constitute an extremely heterogenous group of molecules (Structure and polymerization degree heterogeneity) providing a wide range of potential biological activity. Highest content are found in brown seaweeds, where phlorotannin range from 5-15 per cent of the dried weight. Antioxidant activity of polyphenols extracted from brown and red seaweeds has already been demonstrated by in-vitro assays. However, in contrast to terrestrial plant polyphenols, little is known about the role of phlorotannin in preventing diseases linked to oxidative stress.

15. Minerals

Brown algae, particularly Laminariales are rich in soda, potash, iodine and alginic acid. Ash of dry kelps is a source of soda, which is used in the manufacture of soap, glassware and alum. Bromine is extracted from some red algae such as *Polysiphonia, Rhodymenia.* Many important minerals like copper; iron, zinc, cobalt, vanadium, manganese, boron and chromium are present in high proportions in seaweeds. Hence seaweeds are used as a supplement in stock feed and natural fertilizers. Seaweeds draw from the sea incompletely wealth of mineral elements, macro elements and trace elements. The mineral fraction of some seaweeds accounts for up to 36 per cent of dry matter. Some of the minerals are necessary for our health. The brown seaweeds have traditionally been used for thyroid goiter for their high iodine content. Most of the seaweeds are rich in Ca, P, Mg, Fe, Na, K, CI, Mn, Zn, Cu and Co.

After the death of diatom cells frustules usually dissolve but under certain environmental conditions they may remain intact and accumulate at the bottom of the water body where they occur. Diatomite is insoluble, chemically inert and shows exceptional physical properties that make it suitable for use in many industries. In oil and chemical industries it is used as a filter. It is heat resistant hence used as insulator in boilers and blast furnaces. Because of its inerc nature, diatomaceous earth is used as an absorbent of nitroglycerine, Diatomite is used in the manufacture of car and silver polishing powders, water glass, toothpaste, paints and phonograph records. It is also used as a filter for battery boxes and with Bakelite for switch and fuse boxes.

16. Algae as Bio-fertilizers

Blue-green algae act as nitrogen fixing agent in rice fields. *Anabaena cylindrica.*

Tolypothrix tenuis, Aulaosira fertilissima. Cylindrospermum, Calothrix, Gleoxrichia, Scytonema, Stigonema, Tolypothrix, Oscillatoria princeps, Nostoc commune and many other members of Cyanophyceae have the capability of fixing atmospheric nitrogen. Blue-green algae and *Azolla* constitute a system, which is the main source of algal biofertilixzers in South and South-East Asia. By inoculating rice fields with blue-greens, paddy yield can be increased by 30 per cent.

Larger brown and red algae are used as organic manure. They are usually rich in potassium but have relatively low nitrogen and phosphorus proportions. The problem of fertilizers in developing countries can be solved to a great extent if blue-green algae are used in combination with brown and red algae.

Blue-green algae also help in reclamation of saline and alkaline soils. The growth of blue-green algae in saline and alkaline waterlogged fields results in decrease in pH and increase in phosphorus, nitrogen and organic matter contents of the field and thus converting it into fertile and cultivable land.

17. Seaweed Manure

The earliest record of utilization of seaweed as manure was that of Palladius in 4th century AD as described by Chapman andChapman (1980). It has been found that seaweeds contain many growth promoting hormones such as auxins, cytokinins *etc.* apart from macro and micro-nutrients. Hence, seaweeds can be used as eco-friendly manures either as compost or the extract as liquid seaweed fertilizer (L.S.F.).

18. Seaweed Compost

Basal application of seaweeds as a green manure to coconut and other plantation crops along the Lakshadweep islands and coastal areas of Tamil Nadu and Kerala is still in practice. This treatment improves the water holding capacity of soil, besides supply of micro and macro nutrients upon decomposition. A method for composting seaweeds with cow dung has been described by Thivy (1960). Field experiments have been conducted in the Central Marine Fisheries Research Institute applying seaweed compost to bhendi, sweet potato, tapioca and brinjal plants and high yields were obtained from these vegetable crops.

19. Liquid Seaweed Fertilizer

Now, seaweed extract is made into mineral rich liquid seaweed fertilizer (LSF) and marketed under various trade names. Studies have proved that extracts of *Sargassum wightii, Ulva lactuca* and *Spatoglossum asperum* at 1 per cent strengthshow favorable response on the germination, seedling vigour, fruit settling and on the weight of fruit in crops like groundnut, maize, gingelly and tomato. Liquid seaweed extract was first patented in the year 1912. Another patent was offered in 1962 and exploited by Maxicrop Ltd and marketed as 'Maxi crop'and 'Bio-extract'. When foliar feeding became an orthodox method of plant nutrition in the 1950s 'Marinure', 'SM-3' and 'indent' brands were made in the UK in the 1966s and 'algifert' in Norway.

In India, SPIC is manufacturing and marketing LSF in the name of 'Cytozyme'. The following steps are adopted in the preparation of liquid seaweed fertilizer - thoroughly wash the seaweed to remove sand, debris and other weeds. Dry seaweed and pulverize in a grinder. Soak the seaweed powder in water. Cook the seaweed for 2 hours with water in the ratio of 1:10. Filter and centrifuge the extract. The viscous filtrate is used as 'LSF'. The filtrate is dried at 65-70°C to get dry solid. The dry solid extract is powdered and packed in air-tight bottles. This powder can be used as LSF by making 0.5 per cent to 1.5 per cent (w/v) solution with water. This is used as a foliar spray on green canopy of leafy vegetables and other horticultural crops.

20. Algae as Disposal of Sewage

Oxidative breakdown of sewage is carried out in small tanks and oxygen required by bacteria for this process (oxidation) is obtained from algal growth. Such tanks support luxuriant growth of many unicellular forms, such as *Chlorella, Chlamydomonas, Kuglena, Scenedesmus* and *Pyrobotrys.* Oxygen evolved by these algae during photosynthesis is utilized in oxidation by bacteria. The algae in turn utilize ammonia nitrogen as nutrient.

21. Algae as a Source of Renewable Energy

Several algae (like *Scenedesmus, Chlamydomonas, Dunaliella, Porphyndium* and *Oscillatoria*) possess the enzyme hydrogenase, which can be gainfully employed for the production of hydrogen from water. In blue-green algae such as *Anabaena* nitrogenase enzyme is the chief hydrogen-producing enzyme. Hydrocarbons produced by some algae (*e.g., Botryococcus braunii*) may also serve as liquid or gaseous fuels.

22. Role of Algae in the Origin of Petroleum and Fuel Gas

The fuels such as petroleum and natural gas have their origin in the organic matter in the marine environment. The organic compounds derived from the dead bodies of plants and animals constituting the plankton gradually accumulated at the bottom of the ocean and were buried in the course of time by sedimentary action.

23. Algae as Experimental Material

Algae provide valuable experimental materials for research work in plant physiology, genetics and biochemistry. A lot of researches in genetics and cytology have been carried out of *Acetabularia. Chlorella* has been intensively used for studying the path of carbon during photosynthesis. *Valonia* and *Halicystis* are especially suitable for experiments on membrane permeability.

Fossil Algae

Since almost all algae are soft and delicate, their fossilisation should not be expected. Yet a sizable number of algal fossils have been discovered although fossil algae are imperfectly preserved. Very few algae have been well preserved and have been identified since identification of alga is difficult. Algae fossils have also been imperfectly explored. According to Champman and Champman (1981) the preservation of algae, particulary in the lower geologically state either has not occurred or has not been very good to be identified. Only those algae that deposit lime or silica have been preserved most successfully and therefore these constitute the groups of fossil algae about which we know. Even at the present time fossilisation of algae is taking place.

Table

Sl.No.	Class of Algae	Period in which they Appeared
1.	Cyanophyceae (a) Unicellular (b) Multicellular	Early precambrian
	Oscillatoriaceae	Middle precambrian
2.	Rhodophyceae	Late precambrian
	Chlorophyceae	Late precambrian
	(Chlorococcaces)	Late precambrian
	Phaeophyceae	Late precambrian
	Heterocystous blue green	Late precambrian
3.	Rhodophyceae	Cambrian
4.	Chlorophyceae	Ordovician
5.	Charophyceae	Silurian
6.	Bacillariophyceae	Jurassic
7.	Chrysophyceae	Jurassic
8.	Xanthophyceae	Cretaceous
9.	Chrysophyceae	Cretaceous

Algae have been preserved as impressions, carbon fillows, molds, casts and petrifications. Fossils of algae have been reported from the precambrian period to the recent times. Largest number of algal fossils belong to Cretaceous period although algal fossils have also been reported from Palaeozoic, Triassic, Jurassic and Tertiary periods.

The oldest fossil alga is reported to be a cyanophycean alga - *Archaeosphaeroides barbertonensis.* Blue green algal fossils have been reported from early precambrian era. These fossils are mostly impressions which are globular or thread-like and unicelluar. In the middle precambrian period diversity of fossils increased as coccoid and filamentous blue green algae belonging to Oscillatoriaceae appeared and in late precambrian period the algae which appeared were heterocystous prokaryote and eukaryote. The eukaryote belong to class Chlorophyceae, Phaeophyceae and Rhodophyceae. The following table summarises the algal classes that appeared in different geological period:

A large number of algal fossils have been reported from India as well. These have been reported from Punjab, Bihar, Madhya Pradesh, Rajasthan, Kuch, Kashmir, Kerala, Andaman and Nicobar, Pondichery, Kathiawar, Chhatisgarh, Vindhyans, Indore *etc.* A list of Indian algal fossils is given below:

List of Algal Fossils

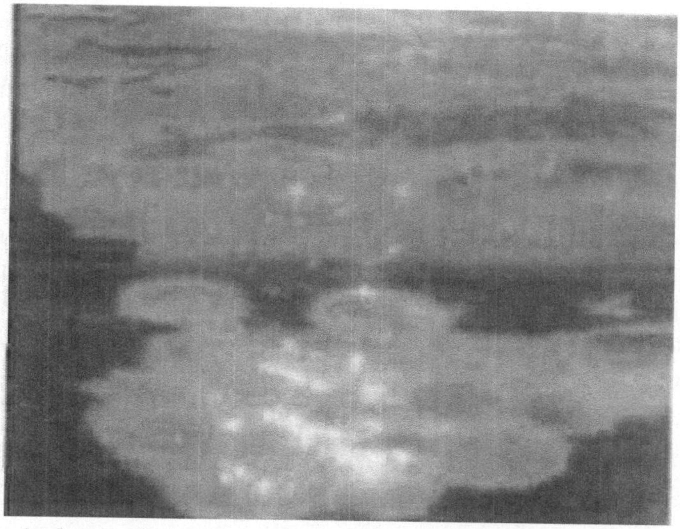

Arcbaeospbaeroldes barbertonensis **Oldest Fossile Alga.**

Sl.No.	*Class*	*Name of the Algal Fossil*
1.	Cyanophyceae	*Aphanocapsa, Synechocystis, Scytonema, Palaeonostoc*
		Archaeonema
2.	Chlorophyceae	*Dissocladella, Dictylospora, Neomeris*
		Piania
		Botryococcus, Palaeochyla, Oedogonium, Ulothrix
		Spirogyrites
3.	Charophyceae	*Chara (Gyrogites), Nitellites*
4.	Bacillariophyceae	Fossil diatoms
5.	Rhodophyceae	*Archaeolithothamnion*
		Lithothamnion, Melobsia
		Distichoplax, Amphira, Corallina, Petrophyton
		Solenopora, Lithophyllum

The study of algal fossils has assumed importance since it was revealed that algal fossils may be indicators of oil deposits. Uses of the study of algal fossils are as follows:

(*i*) Algal fossils may indicate the presence of oil deposits.

(*ii*) Algal fossils have been used to understand the past environment and ecological conditions.

(*iii*) Algal fossils also act as guide fossils since Ordovician of Paleozoic era is also called the Age of Rhodophyta due to dominance of Rhodophycean algae.

(*iv*) Algal fossils are important because they form the limestones. Bricks made of algal limestones are very light and hence used for the construction of high rise buildings.

(v) Diatomaceous earth formed of fossil diatoms are used for various purposes. Alfred Nobel is reported to have used the diatomaceous earth to absorb nitroglycerine from which the Diatomite is made. Diatomaceous earth is also used for insulation of boilers and earth furnaces. In sugar refining process and brewing industry processed diatomaceous earth is used mainly as filtration aid. It is also used as filters for paints, varnishes and paper products. Diatomite is also used for sprinkling to reduce secondary explosions in coal mines.

SELF LEARNING MATERIALS

1. Introduction

- ☆ Evolution of sex in Algae: Originally when sexual reproduction had become established both the gametes were flagellated and of the same size.
- ☆ Morphologially as well as physiologically they were identical. This condition in retained by some of the green and brown algae such as Chlamydomonas, Cladophora and Ectocarpus.
- ☆ In all these forms sexual reproduction is seen its simplest form. It consists in the fusion of similar, motile and naked gametes. They are called the isogametes. They fuse in water.
- ☆ The isogametes are indistinguishable as to sex and are produced in ordinary vegetative (undifferentiated) cells.
- ☆ This kind of sexual reproduction which consist of the fusion of morphologically and physiologically similar gametes is called isogamy.
- ☆ It is the most primitive method of sexual reproduction. It marks the transition from a sexual reproduction to sexual reproduction.

2. Objective: Reproduction in Algae

- ☆ The chief object of reproduction, therefore, is the perpetuation of the species and consequently the increase in number of the individuals of a species. There are two main types of reproduction, a sexual and sexual.

☆ **A sexual reproduction:** It consist of in the separation from the parent of a highly spcialized cell or a group of cells which directly develops into a new individuals resembling of parent.

☆ A sexual reproduction is uniparental and takes place in a variety of ways.

☆ **Spore Formation:** The spores are reproductive units specialized for asexual reproduction. Each unit can grow into a new organism by itself. It involves the sex and there is only one parent. Motile they are called the Zoospore. Non-motile spores are called aplanospores.

☆ **Zoospores:** Formation of Zoospores is the most characteristic method of asexual reproduction in most of the green algae.

3. A Planospores

They are non-motile and constitute a normal means of asexual reproduction in the terrestrial species of the yellow green algae (Vaucheria).

☆ The walls of the aplonospores are thin as compared with the other non-motile spores.

☆ The aplanospores produced in Chloralla have the same distinctive shape as the mother cell before liberation. They are called autospores.

☆ **Hypnospores:** Under certain more stressing environmental conditions the aplanospores in green algae, Secreate thicker walls around them. They are called hypnospores.

☆ **Akinetes:** They are resting cells which mainly serve as means of preservation rather than the multiplication.

☆ The entire protoplast of the vegetative cell rounds off and the original parent cell wall. Considerably thickened.

☆ The akinetes contain abundant reserve food and formed next to a heterocyst or at the end of a trichome in the blue-green algae.

☆ In some they are intercalary in position in green algae. eg.: Pithophora and Cladophora.

3. Content

Isogametes – Morphologically, Physicologically similar gametes – isogamy. Male bearing called antheridia and the female oogonia. Spore formation Zoospore, Aplanospore, Hypnospores, Akinetes, Cysts and statospores. Neutal spores, Monospores, Carpospores, Hormospores, Nanocysts, Pseudohormocyst, Tetraspores.

4. Revision Points

☆ Zoospore are larger and the gametes are smaller.

☆ When division is long continued, it results in the formation of smaller, swarmers, resembling zoospores.

☆ In ulothrix zonata three kinds of Zoospores.

(i) Quardin flagellated macro Zoospores formed in small number per cell.

(ii) Quardin flagellated micro Zoospore fund in large number.

(iii) Biflagellated microzoospores formed in still large number per cell.

☆ The gametes in Ulothrix are also biflagellated and are produced in large numbers.

☆ A sexual reproduction through different spore formation.

☆ Sexual reproduction through a) Isogamy, b) Heterogamy (i) Anisogamy, (ii) Oogamy.

☆ Modes of perennation in Algae: though a) Akinete, b) Secreation of mucilaginous sheath, c) Zygospore, d) Palmellastage, e) Hypnospore, f) Hormospores, g) Tubers.

5. Intext Question

1. What are algae? With the help of suitable examples and illustratios trace the origin and evolution of sexuality in algae.
2. Write an account of origin an development of sex in algae.
3. Describe the various mode of perennaton in fresh water algae studied by you.
4. Describe the evolution of sex in algae and reproduction diversity in different groups of algae.

6. Summary

☆ The evolution of sexual differentiation in the algae has taken place along several independent lines and not in any on phylogenetic line. One of these is volvox series stating with Gonium, passing through pandorina, Eudorina and finally culminating in Volvox.

☆ Ecology of Algae: They are divirsified in variety of environmental adaptation like. Aquatic algae, planktonic algae, Benthic algae, Terrestrial Alga, Lithophytic algae, Holophytic algae, Thermal algae, Snow algae, Aerophytic algae, Epiphytic algae, Endophytic algae, Endozoic algae, Epizoic algae, Parasitic algae.

☆ Algae in soil fertility: Some of the BGA, were deposited in the soil due to aerobic respiration. They enrich soil fertility.

- ☆ The presence of combined nitrogen in the medium inhibits heterocyst development. The formation of heterocyst is inversely related to the amount of nitroten in the medium.
- ☆ The cellular C:N ratio plays an important role in inducing and controlling the differentiation of vegetative cell into heterocyst.

Culture of Algae

- ☆ Culture of algae is important in studying the details of external and internal structures and reproduction of a particular algae or algal group.
- ☆ It is also also important to study the details of their life histories, taxomony, physiology, biochemistry, genetics and molecular studies.

Algal Pigments

- ☆ Chlorophyll a, b, c, d and e of these Chl a is present in all groups of algae.
- ☆ Chlorophyll b is only in Chlorophyceae and Euglenophyceae.
- ☆ Chlorphyll c is found in Phaeophyceae, Cryptophyceae, Baillariophycea and Chrysophyceae
- ☆ Chlorophyll a in red algae only and Chlrorophyll c in certain Xanthophyceae.
- ☆ Phycoerythin, and phycocyanin is found in Cyanophyceae and Rhodophyceae.

Water Bloom

When the algae are abundant in a pond or lake as to give it a distinct colour to the entire body, such as algal growth is called as Water Blooms.

7. Terminal Exercise

Select the Correct Answer

1. Auxospore formation takes place in

   a) Desmids b) Diatoms

   c) Green algae d) Blue green algae

2. A motile flagellated cell that reproduces asexually is called as

   a) Sperm b) Zoospore

   c) Oospore d) Zygospore

3. Chlorophyll e is present in

   a) Chlorophyce b) Xanthophyceae

   c) Phacophyceae d) Rhodophyceae

4. Reproduction through: hormogonium is found in

   a) Spirogyra b) Oscillatoria

   c) Ecotocarpus d) Sargassum

5. Alginic acid derivatives are extracted from

   a) Chlorella b) Spirulina

   c) Laminaria d) Gracilaria

6. Diploid structure in

   a) Tetraspore b) Carpospore

   c) Hypnospore d) Aplanospore

7. Diplobiontic triphasic life cycle is found in

   a) Ulva b) Ectocarpus

   c) Gracilaria d) Stigonema

8. Assignments

1. Give an account of origin and evolution of sex in algae.
2. Describe various modes of reproduction in Algae.
3. Give a detailed account of algal pigments and their significance in algal classification.
4. Trace the origin and evolution of sexuality in green algae. Illustrate your answer with suitable sketches and examples.
5. Write what you know about various types of life cycles in algae.

9. Learning Activities

1. Explain the evolution of thallus in algae given suitable examples from the genera you have studied.
2. Discuss the Economic Importance of Algae.
3. What is mean by culture? Role of Macro and micronutrients in algal culture.

10. Key words

Sex in algae – Alternation of generation in Algae – Algal pigments.

Water Bloom – Fossile algae – Symbiotic algae – Reserve food product in Algae Culture and Cultivation of Algae.

Unit 4

Bryophytes

Introduction and General Characters

1. The bryophytes are a small group of primitive land dwellers. They have a leafy or thalloid, green plant body which is small in stature, rarely more than 5 inches.
2. In their vegetative structure they have become completely adapted to the land habit. However, they still rely upon water for sexual reproduction because the swimming habit is retained by their sperms.
3. The plant body lacks the true roots, stem or leaves. It is relatively simple in the lower forms and still reminds of thf thallus of an alga. It grows prostrate on the ground and is attached tojhe substratum by delicate, unbranched, unicellular hair-like organs called the rhizoids. In the higher bryophytes (mosses) the plant body is erect. It consists of a central axis which bears leaf-like expansions. It is attached to the substratum by branched, multicellular rhizoids.
4. Like the thallophytes the most conspicuous phase in the life cycle is the gametophyte. It is independent and concerned with sexual reproduction.
5. The bryophytes like the thallophytes lack the vascular tissue (xylem and phloem) characteristic of the higher plants.
6. In sexual reproduction they show a marked advance over the thallophytes. It is invariably highly oogamous. The sex organs are jacketed and multicellular. In algae they are non-jacketed and unicellular.

7. Female sex organ in the form of an archegonium appears for the first time in the bryophytes in the plant world.
8. The sperms are biflagellate. Both the flagella are of whiplash type.
9. Fertilization takes place in the presence of water.
10. The fertilized egg is retained within the venter of the archegonium. It neither becomes independent of the parent gametophyte nor passes into the resting period. In both these respects the bryophytes differ from the algae.
11. The zygote undergoes repeated division to form an undifferentiated, multicellular structure called the embryo, ^he first division of the zygote is transverse and the apex of the embryo develops from the outer cell. Such an embryogeny is called exoscopic. It is characteristic of the bryophytes. *No embryo stage exists in the thallophytes.*
12. The venter wall enlarges with the developing embryo to form a protective, multicellular envelope, the calyptra.
13. The embryo by further division and differentiation produces a relatively small spore producing structure which is not independent. It is called the sporogonium (sporophyte). In all bryophytes, the sporophyte is without differentiation into stem and leaves. It is rootless and consists of a foot, a seta and a capsule. In some the seta is absent (*Corsinia*) and rarely both, the foot and the seta (*Riccia*).
14. The sporophyte is thus simpler than the gametophyte and is organically attached to the parent gametophyte throughout its life. It is dependent upon it partially or wholly for its nutrition. In Algae the sporophyte generation when present is always independent of the parent gametophyte.
15. The sporogonium is concerned with the production of wind-disseminated, non-motile, cutinized spores which belong to the category of gonosporesor meiospores.
16. Morphologically the meiospores in a given species are of one kind. Thus the bryophytes, in general, are described as homosporous.
17. Each spore on falling on a suitable soil germinates to give rise to the gametophyte plant either directly or indirectly as a lateral bud from the protonema.
18. The occurrence of heterologous type of alternation of generations is a constant feature of the life cycle of bryophytes. In Algae, when present, it is of homologous type.

Classification, Origin and Evaluation of Bryophytes

The word Bryophyta was coined and used for the first time by Braun (1864) who included algae, fungi, lichens and mosses in this group. In a later system of classification, algae, fungi and lichens were placed in a separate divisior. Thallophyta and liverworts and mosses to Bryophyta. The rank of a division Bryophyta to this well-defined group of plants was first given by Schimper in 1879. Later Eichler in 1883 recognised two subgroups. Hepaticae and tht Musci in the Bryophyta. Since then these two classes remained as separate, major entities within the division. Engler (1892) subdivided each of the two classes into three orders as follows:

Bryophyta

- Class 1. Hepaticae or Liverworts
  - Order 1. Marchantiales
  - Order 2. Jungermanniales
  - Order 3. Anthocerotales
- Class 2. Musci or Mosses
  - Order 1. Sphagnales
  - Order 2. Andreales
  - Order 3. Bryales

This system of classification with slight modification is still followed by many eminent bryologists. Chopra (1981) considered the Bryophyta to be triphyletic. The three sub-divisions proposed by him are Takakiophytina, Hepatophytina, and Muscophytina. The anomalous position of Anthocerotales as an order of the class Hepaticae was pointed out by many investigators very early. The important among them were deBary, ex Janczewski (1872), Leitgeb (1879) and Underwood (1894)^However, it was Howe who in 1899 gave the class status to the order Anthocerotales. He named it Anthocerotes and divided the Bryophyta into three classes: Hepaticae, Anthocerotes and Musci. Many eminent hepaticologists liice Campbell (1918, 1940), Smith (1938, 1955),Takhtajan (1953) followed him and supported his move to divide the Bryophyta into three classes but Smith, Takhtajan and later Wardlaw (1955) and Schuster (1958) called the Anthocerotes a? Anthocerotae. The consensus of opinion at present favours this view and the Bryophyta is classified into three classes, namely, Hepaticae, Anthocerotae and Musci.

Rothmaler in 1951 suggested the following class taxons for the old ones: Hepaticopsida for Hepaticae, Anthocerotopsida for Anthocerotae and Bryopsida for Musci. The new' names suggested by Rothmaler have been recognised by the International Code of Botanical Nomenclature. Proskauer in 1957 suggested the name Anthocerotopsida for Anthoceropsida. The modern bryologists thus classify Bryophyta into the following three classes, namely, Hepaticopsida, Anthocerotopsida and Bryopsida.

Bryophyta

Hepaticopsida
(Hepaticae)
Liverworts
Orders:
1. Takakiales
2. Calobryales
3. Jungermanniales
4. Metzgeriales
5. Marchantiales
6. Sphaerocarpales
7. Monocleales

Anthocerotbpsida
(Anthocerotae)
Hornworts
Order: Anthocerotales

Bryopsida
(Musci)
Mosses
Subclasses:
1. Sphagnidae
   Order: Sphagnales
2. Andreaeidae
   Order: Andreaeales
3. Buxbaumidae
   Order: Buxbaumiales
4. Bryidae
   Orders:
   (i) Fissidentales
   (ii) Dicranales
   (iiI) Pottiales
   (iv) Grimmiales
   (v) Funariales
   (vi) Schistostegales
   (vii) Tetraphidales
   (viii) Eubryales
   (ix) Isobryales
   (x) Hookeriales
   (xi) Hypnobryales
5. Polytrichadae
   Orders:
   (i) Polytrichales
   (ii) Dawsoniales

The above mentioned system of classification has been adopted by Takhtajan (1953), Schuster (1953), Parihar (1965) and Udar (1976). However, Tippo (1942) used the term Atrachaeta for Bryophyta while Takhtajan (1943) and Lam (1948) preferred the use of word Bryopsida for Bryophyta.

The division includes more than 24000 species grouped in approximately 960 species. The science of studv of bryophytes is called *Bryology* and the scientist as *Bryologist.* Salient features of classes Hepaticopsida, Anthocerotopsida and Bryopsida

(a) Hepaticapsida

(i) The plant body thalloid or foliose.

(ii) Rhizoids are present and without septa; usually simple and tuberculate; ventral scales present.

(iii) Cells have chloroplast without pyrenoids.

(iv) Green cells contain simple or compound oil bodies.

(v) Sex organs develop from superficial cells on the dorsal side of the thallus. except when they are terminal in position.

(vi) Sporophyte simple or differentiated into foot, seta and capsule.

(vii) Sporogenous cells develop from endothecium.

(viii) Columella is absent, elaters maybe present.

(ix) Sporophyte completely dependent upon the sporophyte for nutrition.

(x) Dehiscence of sporogonium is irregular or regular.

(b) Anthocerotopsida

(i) The plant body is thalloid and dorsiventrally flattened.

(ii) Rhizoids only simple without septa, ventral scales absent.

(iii) No tissue differentiation in thallus;cells bear a large chloroplast with a conspicuous pyrenoid.

(iv) Sex organs embedded in the thallus tissue.

(v) Antheridia develop from the hypodermal cells of the thallus on the dorsal side; located in the antheridial chambers singly or in groups.

(vi) Archegonia sunken on the dorsal side of the thallus, develop from superficial cells.

(vii) Sporogonium elongated and cylindrical, arise from the dorsal side of the thallus.

(viii) Sporogonium has foot, a meristematic region and capsule, meistem is intercalary in position and due to this sporogonium continues its growth throughout the growing season.

(ix) Capsule wall has chlorophyll and stomata

(x) Columella and pseudo-elaters are present.

(xi) Sporogenous tissue develops from amphithecium.

(c) Bryopsida

(i) The primary gametophyte consists of prostrate, filamentous or thalloid protonema and erect gametophore.

(ii) The adult gametophyte consists of 'stem', spirally arranged 'leaves' and the sex organs at its apical portion.

(iii) Rhizoids multicellular branched with oblique septa.

(iv) Sex organs develop from superficial cells.

(v) Sporophyte differentiated into foot, seta and capsule.

(vi) Capsule wall has stomata.

(vii) Sporogenous mass develops mostly from outer layer of endothecium which in addition forms columella.

(viii) Elaters are absent.

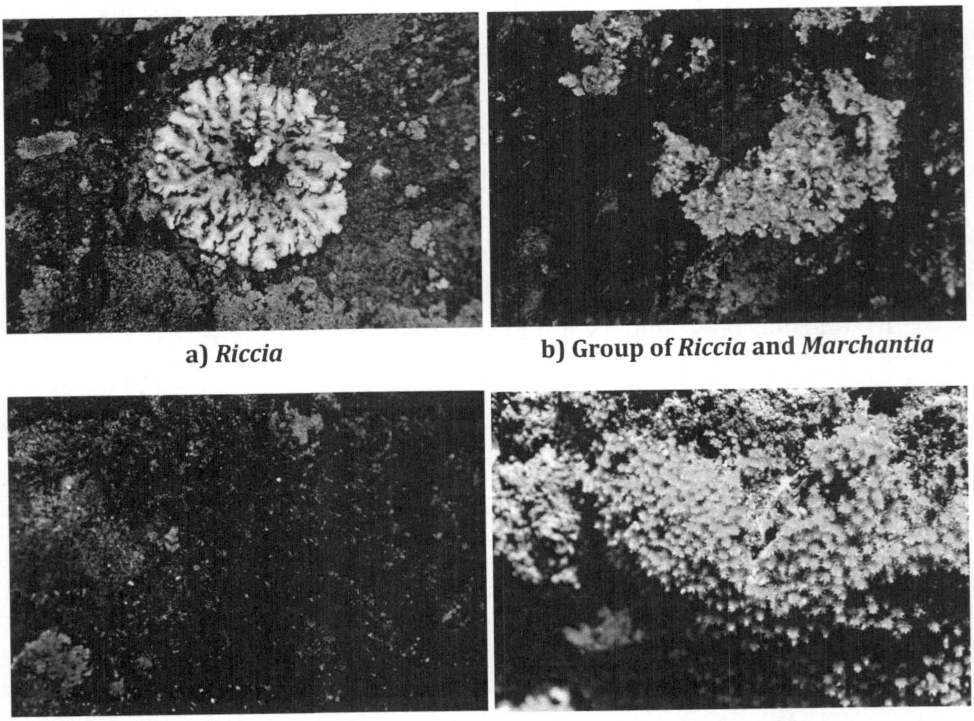

a) *Riccia* **b) Group of *Riccia* and *Marchantia***

c) *Targenia* and *reboulia* **d) *Briyum* and *Mosses***

Figure 4.1: Natural Habitat of Bryophytes

Alternation of Generations (Figure 4.2)

The life cycle of bryophytes is very interesting. It is split up into two phases, each represented by a separate adult. Thus in a single life cycle there occur two distinct multicellular, vegetative individuals. One of these is the green **thalloid** (liverworts) or **leafy** (mosses) plant. The other is the **sporogonium.** The green individual is the most conspicuous of the two. It is an independent plant.

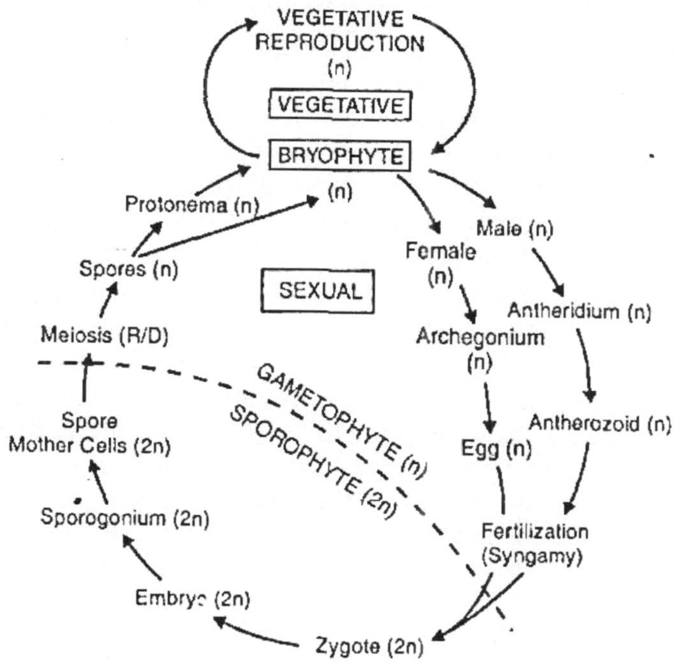

Figure 4.2: Bryophytes, Graphic Representation of the Sexual Life Cycle.

It is haploid and bears the sex organs (antheridia and archegonia) which produce the gametes (sperms and eggs). As it bears the gametes the green, independent individual is called the gametophyte. It is concerned with sexual reproduction. The gametophyte plant along with the structures produced by it constitutes the gametophyte generation, In the life cycle it starts with the formation of meiospores and consists of the green individual and the sex organs. The last structures formed during this phase are the gametes.The gametes fuse to form the zygote.

The zygote, on germination, does not produce the gametophyte plant. It undergoes segmentation to form ai. embryo. The embryo by further segmentation and differentiation gives rise to the second adult called the sporogonium. It remains diploid and is usually differentiated into foot, seta and capsule, in due course of time, the diploid spore mother cells produced in the capsule give rise to haploid spores (meiospores) by meiosis. As the sporogonium is concerned with the production of spores it is called the sporophyte. The zygote, the embryo and the sporogonium together constitute the sporophyte generation. It is less conspicuous. Moreover, it is dependent for its nutrition wholly or partially on the gametophyte plant to which it is attached organically throughout its life. It starts with the zygote and ends with the formation of spore mother cells in the capsule. The meiospores are the pioneer structures of the next gametophyte generation. On germination each spore produces a gametophyte and not a sporophyte plant.

From the account given above it is evident that on germination the reproductive cells of one generation give rise to the alternate generation in the life cycle. The two generations thus regularly alternate with each other in a single life cycle. This biological phenomenon is called alternation of generations. It is defined as the alternation in the life cycle of two distinct vegetative individuals with different functions. Cycles of this type characterized by alternation of generations and sporogenic meiosis are termed diplohaplontic life cycles.

In the bryophytes alternation of generations becomes as integral part of the life cycle. It is a constant feature of all the species because the zygote invariably develops into a distinct sporophyte. This may be due to the fact that the egg is retained in the tissue of the gametophyte. Fertilisation takes place *in situ.* As a result the zygote has a better chance for growth. Moreover, the alternating individuals in the life cycle of the bryophytes are morphologically dissimilar. They differ not only in their structure but also in their physiology (nutrition). This kind of alternation of generations in which the alternating individuals are dissimilar is called heterologous or heteromorphic. The structural and nutritional difference in the two generations may be an evolutionary sequence due to the different modes of life. The gametophyte is independent. The sporophyte is not. It is attached and generally dependent wholly or partially on the gametophyte. The attached condition of sporophyte can lead to dissimilarity. This Viewpoint is supported by an analogy from the red algae. The attached carposporophyte of *Polysiphonia* is quite different from its free living gametophyte. On the other hand, the free living tetrasporophyte is identical to the gametophyte.

In Algae alternation of generations is not a constant feature of all the species. It occurs in isolated instances. The vegetative sporophyte is generally absent. It is represented only by the single-celied zygote. The zygote is liberated in Algae. Being independent it lacks enough resources to develop into a distinct sporophyte. Moreover, the sporophyte in Algae when developed is always independent of the gametophyte plant. In cases where there is alternation of generations, the two alternating individuals in the life cycle generally resemble each other. Morphologically, they are identical. This type of alternation of generations in which both the alternating generations are structurally similar is called homologous or isomorphic.

Besides food, the two most important biological needs of every living plant species are:

1. Production of new individuals. Really new individuals are produced by the fusion of male and female gametes from different plants. It is called cross-fertilization.

2. Their dispersal to suitable places where they can find their food without much competition.

In Algae both these problems are solved by the agency of water. The gametes swim through water to fuse. The resultant zygotes are dispersed by the water currents. This accounts for the absence of structural differences between the two alternating plants in the life cycle. There is thus no division of labour. Hence, the alternating generations are similar.

Among the land dwellers such as the bryophytes the two probelms of cross-fertilization and dispersal are solved on the basis of division of labour. Each generation caters to and is suitably adapted to solve one of the these problems. The gametophyte is concerned with sexual reproduction. It thus caters to the need of cross fertilization. The sporophyte functions in the multiplication of beneficial effects of a single sexual union. It produces minute spores which are distributed by the wind. The division of labour among the two generations thus accounts for the structural difference between them in the byrophytes.

Alternation of chromosome number in the life cycle. Each cell of the gametophyte generation has a certain number of chromosomes in its nucleus. It is represented by the symbol *n.* The numerical value of *n* differs in the different species. Each gamete thus has *n* chromosomes in its nucleus. It is the basic number. At the time of fertilisation the nuclei of the sperm and egg fuse. *The, paternal* and the *maternal* chromosomes however, do not fuse. They simply become closely associated in pairs in the fusion nucleus. The fusion nucleus in the zygote now has *n* plus nor 2 *n* chromosomes. The single number *n* is called *haploid*or *monoploid*'number. All structures with *n* chromosomes are thus *haploid.* On this basis the gametophyte generation may be called the *haploid generation* in the life cycle. The 2 *n* number is called *diploid.* The zygote with 2 *n* number of chromosomes thus is a diploid structure. The sporogonlum which is developed from the zygote by repeated mitosis is a diploid individual in the life cycle. All the cells of the sporophyte generation (sporogonium) up to and including the spore mother cells are diploid. At the time of *fertilization* the haploid generation switches on to the diploid. With *meiosis* the diploid generation switches back to haploid. The *alternation of generations* is thus accompanied by the alternation of chromosome numbers from the haploid to the diploid and back to the haploid condition.

Origin of alternation of generations. It is still a live question. Two theories, namely, antithetic and homologous have been proposed to explain how alternation of generations originated.

Antithetic theory. It was the first to be proposed. On the basis of this theory the gametophyte or sexual plant represents the original generation. The sporophyte or the non-sexual organism is a new and different phase evolved by progressive elaboration of the diploid zygote of some algal ancestor like *Coleochaete.* It is interpolated in the life cycle of the gametophyte of primitive land plants between the two crucial points, fertilisation and meiosis in response to a life in a drier environment. The factors which caused its origin are prompt germination of the zygote accompanied by delayed meiosis.

The result is the production of a small sporophyte of *Riccia* consisting simply of a spore case. With further elaboration and increased sterilisation of the spore producing tissue a larger sporophyte with differentiation into a foot, a seta, and a capsule is finally evolved.

Homologous theory. On the basis of this theory the sporophyte is simply a modification of the gametophyte and not a new generation evolved in response to a life in a drier environment. The advocates of this theory point to the fact that among the green algae the gametophyte plant reproduces by both the methods of reproduction. It bears spores and also the gametes. In the course of evolutionary sequence these two functions became separated in two distinct individuals. One of these produced the spores and the other produced gametes. The former came to be known as the sporophyte and the latter gametophyte. These two individuals occur regularly one after the other in the life cycle. The occurrence of two similar generations in most of the algae furnishes a strong argument in favour of this theory. The phenomenon of apospory and apogamy is another. Both the individuals in primitive land plants were photosynthetic and free living. Gradually, the sporophyte became attached to and partly parasitic on the gametophyte. Consequently, it became reduced.

Inter-relationship of Bryophytes

Bryophytes are Descended from Pteridophytes

According to another school of thought, the Bryophytes are descended from Pteridophytes by means of reduction. The various theories which support this view are called retrogressive evolutionary theories. This hypothesis is supported by Kidston and Lang (1917-1921) Kashyap (1919), Haskell (1949), Christensen (1954) and Proskauer (1960).

Thanks to the discovery of *Rhyniu* and *Horneophyton* like plants by Kidston and Lang (1917-1921) from the middle Devonian period, which showed affinity of Psilophytales with the sporophytes of Anthocerotales, Sphagnales and Andreaeales, Proskauer (1960) emphasized the similarity which the exists between the sporophyte of Psilophytales and Anthocerotales. He also states that in *Dendroceros crispus,* the outermost layer of collumella and the inner most layer of the jacket showed all the essential features of 'tapetum'. He interpreted that atleast the Anthocerotales might have originated from Psilophytales.

Haskell (1949) also.supports Pteridophytic ancestry of Bryophytes. He states that the origin of the Bryophyta from an algal ancestry is open to question and it may represent a group originating from Psilophytalean ancestry, following reduction due to their habitat.

Christensen (1954) gives three sources of information for this origin from Pteridophytes.

1. From the gametophytes and sporophytes bearing leaves.
2. From leafless sporophytes.
3. From different groups of Pteridophytes.

Kashyap (1919) postulated that atleast the Hepaticopsida might have been derived from Psilophytales. In the absence of concrete documentary fossil proof, the final decision and judgement on the origin of land plants is suspicious and it is kept on continuously.

Nature of Primitive Gametophyte

Among Bryophytes, how evolution is taking place? There are two widely divergent theories regarding the nature of primitive gametophytes of Bryophytes. According to one theory, the primitive gametophyte was as erect leafy shoot, radial in its symmetry. The following workers supported this theory: Wettstein (1903-1908), Church (1919), Evans (1939), Kashyap, Goebel and Mehra (1957).

According to the second theory "The primitive gametophyte was a simple, dorsiventral, prostrate thallus which was simple in its external form as well as histological structure". Cavers (1910) and Campbell (1891, 1918, 1936 and 1940) are the main adherents of this view.

First Theory: The Primitive Gametophyte was an Erect Leafy Shoot, Radial and its Symmetry

According to Wettstein, the primitive gametophyte was more or less like normal plants having well developed rhizoidal system, axis like and leaf like structures. The gametophytes are erect, leafy, and radial. This condition prevails in plants like *Polytrichum* and *Haplomitrium.* The leaves are arranged spirally. From this erect form, gradually prostrate form appears and the course of evolution can be traced as follows:

The spiral arrangement of leaves were replaced by three ranked arrangement of leaves as in *Herbertia.* Owing to the feeble rhizoidal development, the erect gametophytes become a prostrate form. It was changed into a dorsiventral structure. The leaves were unequal in size due to the dorsiventral nature of gametophyte. The dorsal leaves are larger as they are towards light. The ventral leaves are smaller as they are turned away from the light source: This phenomenon is called anisophylly. This type of anisophylly is found in *Porella* sp.

According to Evans (1939), from this anisophyllic condition, the ventral leaves are gradually reduced and eliminated as in *Lepidozia* and *Lejeunea.* In the case of *Plagiochila* and *Cephalozia* the ventral leaves are vestigial (like Papillae). In *Radula,* the ventral leaves completely disappear. After the disappearance of the ventral leaves, the dorsal leaves fuse (*Schiffneria*). In many cases, the sexual branches retain the features of typical, radial erect leafy shoot, which thus approaches, more towards, a thalloid nature.

According to Goebel, *Monoselenium* of Marchantiales has a very simple thallus without air chambers. In the sporogonium (Capsule), elaters are absent. The wall cells lack spiral thickening.

Kashyap, based on his studies on Indian Liverworts, also supports this theory. He traces the retrogressive evolution among the Marchantiales, on the basis of following points:

a. The Loss of Assimilatory Filaments

The plants like *Marchantia* and *Preissia quadrata* grow on moist soil; they show well developed ventilating systems. They have well developed air chambers, air pore and assimilatory filaments. In *Conocephalum conicum,* the air chambers are still reduced and the assimilatory filaments are also very few. In *Wiesnerella denudata,* still further simplication has taken place. Assimilatory filaments are few, one or two celled in height. In *Dumortiera* there is complete elimination of air chambers.

b. Simplification of Air Pore

In *Marchantia* and *Preissia,* a complex form of air pores are seen. There are definite guard cells surrounding each airpore. In *Reboulia,* there is a very simple airpore. In *Riccia,* the most primitive form of airpore is present.

c. Archegoniophore

There is a gradual shifting of the terminal stalk of spherical, erect branches bearing sex organs to the dorsal position by the continued growth of the thallus and gradual elimination of the stalk. In *Marchantia,* the archegoniophore is terminal. In *Plagiochila,* the terminal archegoniophore becomes shifted to the dorsal side of the thallus by the continuous growth of the thallus. In *Reboulia* and *Corsinia* the female receptacles become sessile (without any stalk). Harris (1938 and 1939), on the basis of his discovery on *Naiadita,* a fossil Hepaticopsida, suggests that the primitive type of gametophyte was only an erect leafy form. Mehra (1957) stated that the primitive gametophytes of Bryophytes were like *Stephensoniella brevipedunculata* and *Asterella reticulata.*

The Second Theory

According to this theory, the primitive gametophyte was a simple, dorsiventral, prostrate thallus, which was simple in its external form as well as histological structure. Cavers (1910) and Campbell (1891, 1918, 1936 and 1940) supported this theory. According to Cavers, the primitive gametophyte was like that of *Sphaerocarpos,* and primitive forms, evolution proceeded in two different lines. One line leads to thalloid Marchantiales and the other line leads to leafy Jungermanniales and Calobryales. In Marchantiales, the members have a morphological simplicity but anatomically with complex well developed air chambers, assimiiatory filaments and air pores. But in the Jungermanniales, we find anatomical simplicity and morphological specialization. They have axis-like, leaf-like and root-like structures.

SELF LEARNING MATERIALS

1. Introduction

- ☆ It is represented by the Swamps and the areas where water and land meet. It may will be called the amphibious zone. Imlabing the amphibious zone are the mosses liverworts and hornworts which collectively constitute a group of non-vascular land plants called the bryophytes.
- ☆ Bryophytes thus include archegoniate, atracheate simple cryptogames in which zygote divided by mitosis and forms embryo. Embryo – the sporophytic phase depends upon gametophyte for nutrition and support. These are amphibians of the plant Kingdom.
- ☆ In many liverworts there are numerous pores on the upper surface of the thallus. These are called the air pores. They facilate gaseous echange between the atmospheric air and the interior of the thallus.
- ☆ The fertilized egg is retained within the archegonium. Here it obtains food and water from the parent plant and is protected from drying as it develops into an embryo. The adaptation is essential for the survival of the land plants. It ensures nursing of the young embryo and its protection against mechanical injury.
- ☆ The thick-walled, wind-disseminated spores and the primitive vascular system in the form of a conducting strand are other adaptations to land habit.
- ☆ The Bryophytes are a small group of most primitive land dwellers. The bryophytes number about 24,000 sps which are grouped under nearly 960 genera. All of them are small and in conspicuous plants ranging in size from about a millimeter in length to 30 centimetres or more.

2. Objective

- ☆ Amphibians *i.e.*, grow in moist shady place as well as in water.
- ☆ Tissue differentiation well developed.
- ☆ Sexual reproduction is Oogamous only.
- ☆ Sex organs covered by a sterile jacket.
- ☆ Zygote remain closed in the archegonium.
- ☆ Embryo produced from zygoe.
- ☆ Sporophyte dependent upon gamotophyte.
- ☆ Sporophyte developed and differentiated into Foot, Seta, Capsule.
- ☆ The occurrence of heterologous type of altenration of generations in the life cycle of Bryophytes.

3. Content

- ☆ The plant body thalloid or foliose.
- ☆ Cells have chloroplast without pyrenoids.
- ☆ Green cells contain simple or compound oil bodies.
- ☆ Sporophyte simple or differentiated into food, seta, capsules.
- ☆ Sporogenous cells develop from endothecium.
- ☆ Sporophyte completely depend upon the sporophyte for nutrition.
- ☆ Dehiscence of sporangium is regular or regular.

4. Revision Points

A) *Hepaticopsida*

- ☆ Plant body is either leafy or thalloid.
- ☆ Rhizoids are present and without septa; usually simple and tuberculate; ventral scale present.
- ☆ Sex organs develop four superficial cells on the dorsal, side of the thallus except when they are terminal in position.
- ☆ Sprogenous cells develop from endothecium.

B) *Anthocerotopsida*

- ☆ The plant body is thalloid and dorsi ventrally flattened.
- ☆ Rhizoids only simple withot septa, ventral scales absent.
- ☆ Sex organs embedded in the thallus tissue.
- ☆ Sporogeneous tissue develops from amphithecium.

C) *Bryppsida*

- ☆ Filamentous or thalloid protonema and erect gametophyte.
- ☆ Rhizoids multicellular branched with oblique septa.
- ☆ Sex organs develop from superficial cell.
- ☆ Sporogeneous mass develop mostly from outer layer of endothecium which in addition from Columella.

5. Intext Question

1. Briefly describe the chief means of perennation in liverworts.
2. The plant body of some Bryophytes are Hepatic. Why?
3. How do the sporophytes of liverworts take their nourishment?
4. Describe the various methods of vegetative reproduction in Hepaticae.
5. Briefly describe the salient features and classification of liverworts.

6. Summary

- ☆ Thallus – like plant body – lack of vascular tissue – Absence of roots – Autotropic mode of nutrition.
- ☆ The early stages of development in the gametopyte of many bryophytes are green filaments. Which strikingly resemble the filamenteous thallus of green algae.
- ☆ The carbohydrate food reserve in the Bryohytes is true starch, which contains a mixture of two kinds of glucose macromoleculars, amylose and amylopectin.
- ☆ Condensation theory: Mehra and Vasisht (1950) suggested the derviation of the thalloid structure in the liverworts form the foliase forms by condensation.
- ☆ Later in 1957 Mehra addressed evidence in support of a real phylogenetic connection between the Jungermanniales and the Marchantiales.
- ☆ He propounded a theory explaining the origin of the Marchantiaceous thallus from the foliose Junger mammiales by "compaction condensation and fusion of leaves"

7. Terminal Exercise

1. Arrange the following in Sporangum of Marchantia from outside to inside:

   a) Capsule b) Perigynium

   c) Calyptra d) Elaters

2. Which of the following is not found in Marchantia?

   a) Perigynium b) Perichaetium

   c) Calyptra d) Paraplysis

3. Describe vegetative reproduction in Riccia and Marchantia.
4. What is perianth? Name a few liverworts which show this structure.
5. Define alternation of gemeration and Explain it with reference to the life cycle of Marchantia.

8. Assignments

1. With neat labeled sketches describe antheridiophore and archegoniophore of Marchantia.
2. Give a comparative account of the gamatophytes of Riccia, Marchantia and Anthoceras.
3. Draw diagram antheridia and archegonia of Marchantia and Anthoceros and give structural differences in the sex organs of two plants.

9. Learning Activities

- ☆ Give a comparative account of the structures of gamotophytes of Riccia and Marchantia.
- ☆ With the neat labelled sketches describe antheridiophore and archegoniophore of Marchantia.
- ☆ Draw neat and labelled diagram of the following: a) L.S. Sporophyte of Marchntia, b) Tuberculate rhizoid, c) Single Gemma.

10. Key words

Smooth walled and tuberculate rhizoids, Calyptra, Operurlum, gametophyte, Sporophyte, Elaterophore, Involucre, Jacket layer, Protonema, heterologous type.

Unit 5

Range of Variations in Structure of Gametophytes of Byrophytes

As it is, this lesson comprises thalloid forms, leafy Jungermanniales and Mosses which are included in the syllabus. Thalloid forms: Riccia, Targionia, Dumortiera, Reboulia, Marchantia, Pallavicinia, Riccardia and Anthoceros.

Jungermanniales: *Porella*

Mosses: Sphagnum, Bryum and Pogonatum.

Spore is the first cell of gametophytic generation. Spore falls on suitable substratum develops into a gametophyte. The gametophyte is prostrate, thalloid, dorsiventrally flattened and dicotomously branched in the thalloid forms, prostrate and leafy axis in Jungermanniales (*Porella*) and Prostrate underground rhizome with erect leafy aerial axis in Mosses.

Ricciaceae

Ricciaceae differ from other Marchantiales in having sex organs borne in a sagittal strip extending the entire length of the thallus. Sporophyte consists solely of a capsule. This family includes about. 140 species.' There are three important genera in this family. They are *Tesselina* (*Oxymitra*), *Ricciocarpus* and *Riccia.*

Riccia discolor is a Terrestrial Form

This genus includes about 138 species. The gametophyte is very small, green, dichotomously branched growing as a colony and arranged in the form of a rosette The gametophyte oh the dorsal side has a midrib and a longitudinal groove which end in a depression called apical notch, where the growing point is situated. On the ventral side, along the midrib, there are rhizoids and scales. The rhizoids are of two types, namely smooth-walled and pegged rhizoids. The pegged rhizoids are otherwise called tuberculate rhizoids. The *scales* are multicellular, plate-like structure, but single cell in thickness. Usually scales are arranged in two rows. The gametophyte may be monoecious or dioecious.

Anatomy of the Thallus

The thallus is clearly differentiated into a dorsal side and ventral side.

Dorsal Side of the Thallus

There is ill-defined upper epidermis. It is dis-continuous, due to the presence of air pores. From the floor of the ventral surface, there are many rows of cells. (Figure 5-1)

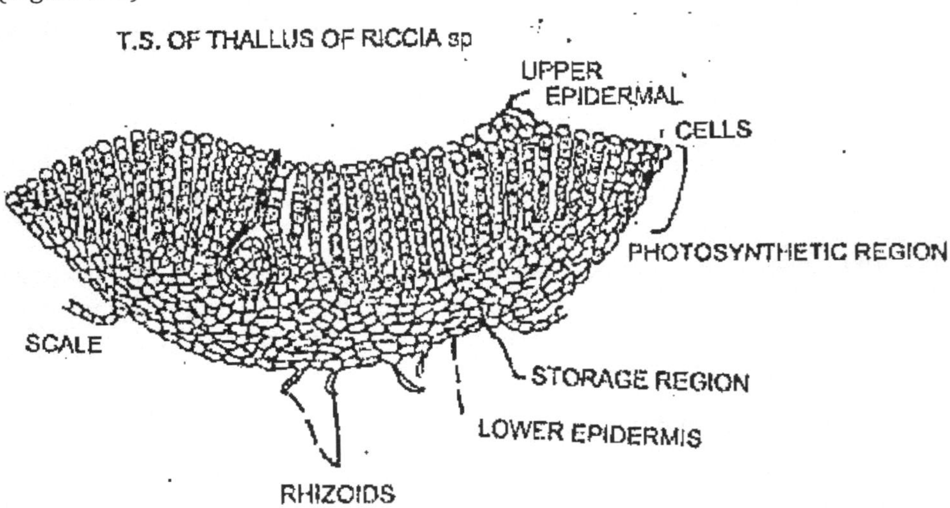

Figure 5.1

Each row may consist of 4-8 cells in height. These cells possess chlorophyll pigment. Hence, this region is called assimilatory region. In between the rows of cells, there are air spaces or air chambers. Each air chambers communicates with atmosphere through a minute opening called air pore. Topmost or terminal cell of each row is devoice of chlorophyll pigments and they are slightly larger than the rest of the cells of the row and these cells constitute 'upper epidermis'.

Ventral Side

Ventral side of the thallus consists of thin-walled, parenchymatous cells, and they are arranged without any inter-cellular spaces. Water and mineral salts and food materials are stored here. Lower epidermis is continuous and single cell in thickness. From this, rhizoids and scales arise. There are two types of rhizoids, namely smooth-walled and pegged rhizoids and they are meant for absorption and anchorage.

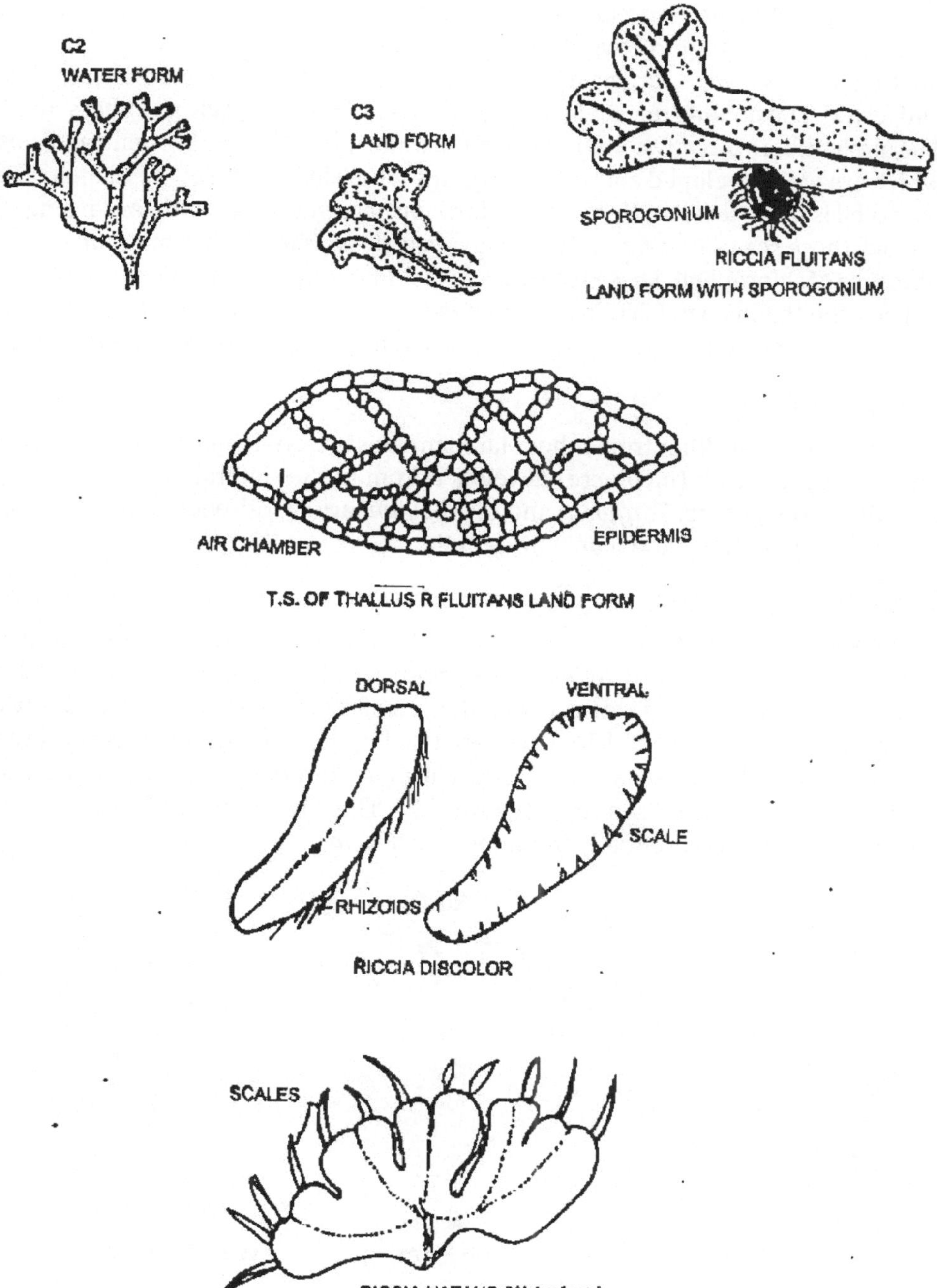

Figure 5.2

Note: *Riccia* has a very poor ventilating system. Both the air chambers and air pores are very simple and primitive. *Riccia fluitans, R.natans* (*Ricciocarpos natans*) are aquatic forms, but they are capable of growing in terrestrial habitat.

The gametophyte is green, dorsiventrally differentiated with marginal lobes. The gametophytes are of two types, namely floating form and land form. The floating form is always larger in size and sterile. The land form is smaller in size and fertile. The floating forms show large number of polygonal air chambers in T.S. The airchambers directly communicate with atmosphere. They have ventral scales which are well developed and scattered in the whole lower side of the thallus. The basal cell is meristematic. Hence, they show long continuous growth. According to Goebel, these scales in *R. natans* are to increase the surface for assimilation due to presence of chlorophyll. These scales are sword-like, long.and hang down obviously to keep the thallus balanced in normal position. In the land forms, the scales are much less developed and serve as protective organ of the growing point. (Figure 5.2)

Targioniaceae

Targioniaceae differ from other Marchantiales in having the female receptacle enclosed by a sheath (involucre) developed.from tissues lateral to it. This family contains three genera. *Targionia* and *Cyathodium* are world-wide in distribution and *Aiichinsoniella* is known only from India.

Targionia hypophylla

Plants are usually dioecious. The gametophyte is green, dichotomously branched and dorsiventrally differentiated, with aeriolae. On the ventral side, scales and rhizoids (pegged and smooth-walled) occur. Scales are arranged in two rows, one on either side of the midrib. There are two large involucres which protect the growing apex. This is the characteristic feature of *Targionia.* The sex organs are borne on special ventral branches (Figure 5.3). The development of antheridium and archegonium is essentially the same as in *Riccia.*

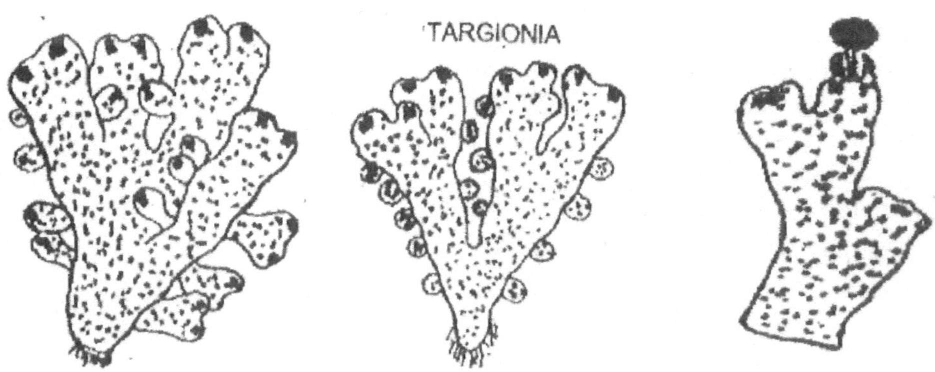

Female Plant **Male Plant** **With Sporophyte**

Figure 5.3

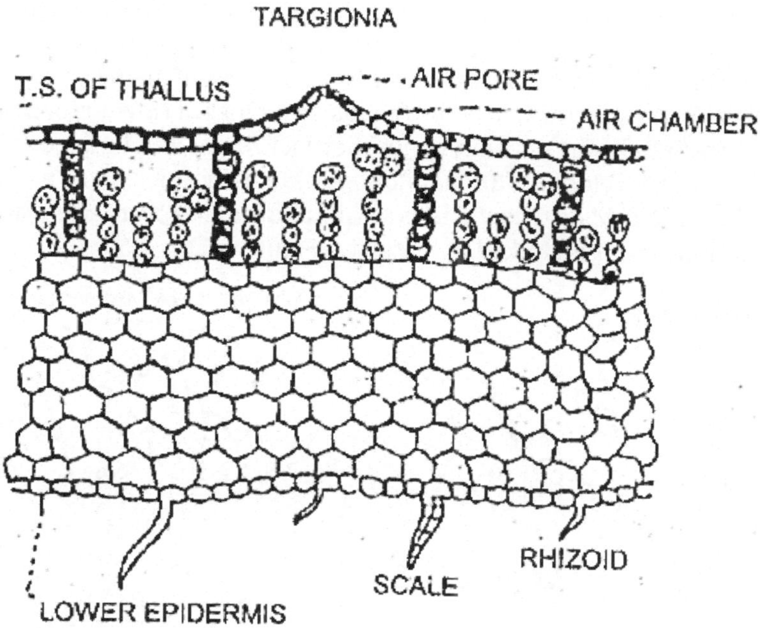

Figure 5.4

Dorsal Side

The upper epidermis is discontinuous by the presence of air pores. The epidermal cells are larger, thickened and are characteristics of *Targionia.* The air pores are simple. Just below the epidermis, there is a single row of air chambers. From the floor of the ventral surface, unbranched assimilatory filaments, of 3 or 4 cells in height are found. In each assimilatory filament, the terminal cell is comparatively larger than the rest of the cellar

Ventral Side

The ventral side is very broad and storage in function. In this region cells are thin walled, parenchymatous and arranged without any inter-cellular spaces. These cells contain starch. Some of the cells posses mucilage or oil droplets. Lower epidermis is continuous and single cell in thickness. From this two types of rhizoids namely smooth walled and tuberculate rhizoids and two rows of scales arise on either side of the midrib (Figure 5.4).

Marchantiaceae: This family includes 23 genera and 250 species, and comprises all the genera in which archegonia are borne on special stalked, vertical branches (archegoniophores or carpocephalum). This family has many genera. The chief among them are *Dumortiera, Roboulia, Plagiochasma, Lunularia* and *Marchantia.*

Dumortiera hirsuta: The thallus is very broad, green, dichotomously branched and dorsiventrally differentiated. The margin of the thallus shows many lobes. Midrib is prominent. Air chambers and air pores are absent on the dorsal surface of J the thallus. On the ventral side of the thallus, two types of rhizoids occur. Most off the plants are found near streams, hence they lack scales, but in terrestrial forms, scales are poorly developed. The gametophyte is usually dioecious. The antheridia occur on the margin. Male receptacle' is sessile and female receptacle is also sessile in the beginning but it is stalked at maturity (Figure 5.5).

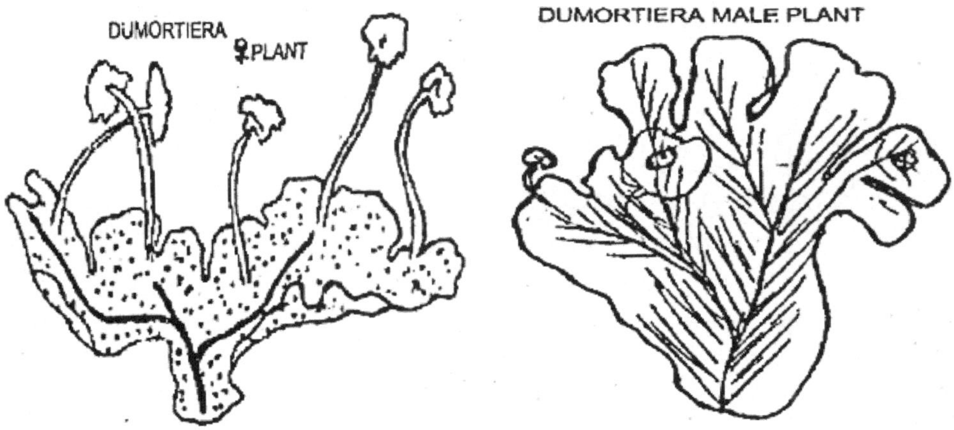

Figure 5.5

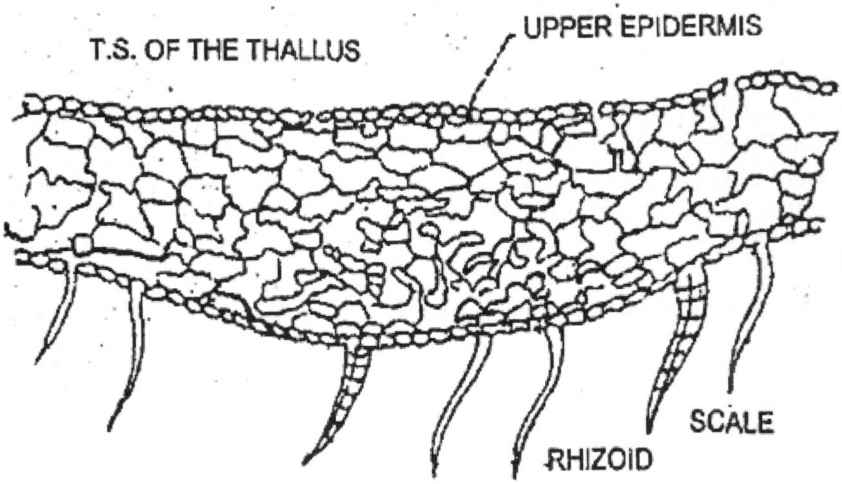

Figure 5.6

Thallus of *Dumortiera* is made up of thin-walled, parenchymatous cells without any intercellular spaces. The cells have chlorophyll pigments and starch grains. Upper and lower epidermises are continuous. From the lower epidermis, Scales and two types of rhizoids (smooth walled and pegged) arise (Figure 5.6).

Reboulia hemispherica **Reddi**

The gametophyte is monoecious or dioecious. It is green, dichotomously branched, dorsiventrally differentiated. Scales are arranged in two definite rows, one on either side of the midrib. Smooth walled and pegged rhizoids are present. Male receptacle is sessile, while the female receptacle is stalked. The stalk has a single rhizoidal furrow (Figure 5.7). Anatomy of the thallus of *Reboulia* shows simple air pore and many layers of air chambers. There are no assimilatory filaments, but instead the cells the wall of the air chambers have chlorophyll pigments. The air chambers are very broad. Ventral surface of the thallus has scales and two types of rhizoids (Figure 5.8).

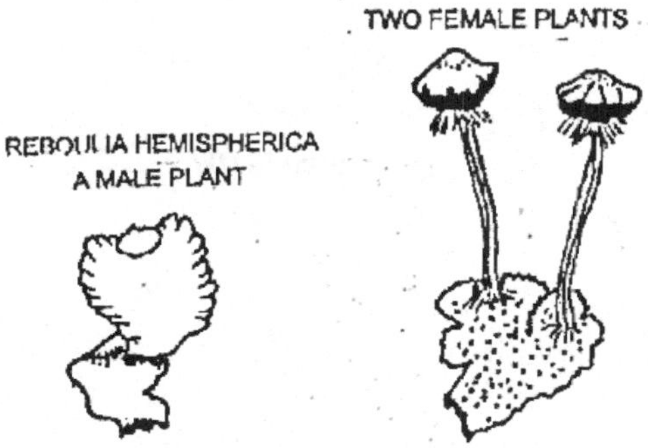

Figure 5.7

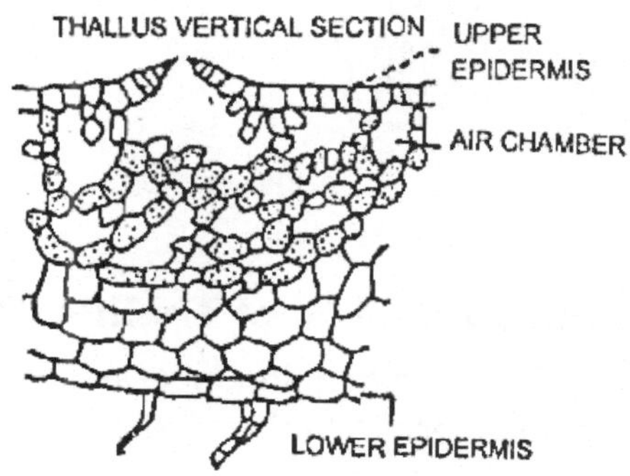

Figure 5.8

Marchantia: This genus has about 65 species.

Marchantia polymorpha **L.**

The gametophyte of Marchantia is dioecious, green, dichotomously branched, dorsiventrally flattened and differentiated. On the dorsal surface of the thallus, there is a definite midrib. On the dorsal side there are gemma cups. The gemmae, which are found in the gemma cups are meant for asexual reproduction. On the dorsal side of the thallus, there are many polygonal areas. Each area has a pore at the centre. Each polygonal area represents the air chamber below it. Each pore represents a respective the air chamber below it. On the ventral side of the thallus, over the midrib, there are smooth-walled and pegged rhizoids and scales. Scales are arranged in two rows. Some scales have appendages. Gametophyte is dioecious. Sex organs are elevated from the thallus to a special reproductive branch called antheridial disc and archegonial disc. Both the gametophytes can be differentiated externally. The archegonial disc is also called carpocephalum. It has 8 or 9 rays. The antheridial disc is an 8-lobed structure. The stalk of the reproductive branches show two rhizoidal furrows (Figure 5.9).

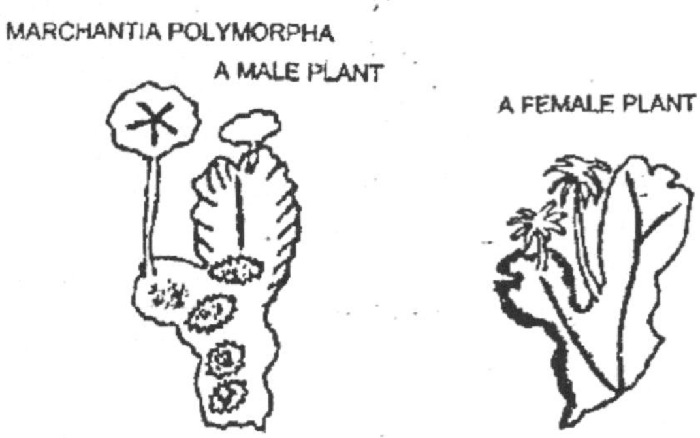

Figure 5.9

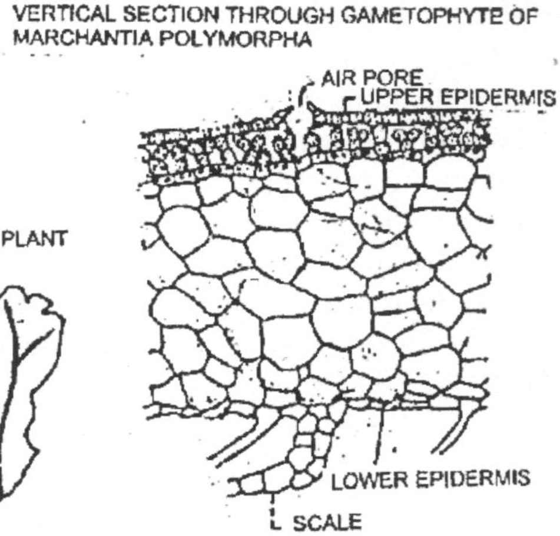

Figure 5.10

Marchantia T.S. of Thallus

There is a definite upper epidermis. Many air pores are present. Below the upper epidermis, there is a single row of air chambers. In each air chamber, from the floor of the ventral surface, branched or unbranched assimilatory filaments of a few cells in height are present. Each air chamber communicates outside through an air pore. Air pore is of a complex type here. Each air pore is surrounded by many guard cells, arranged in successive rings. An air pore is surrounded by 4-5 tiers of cells and each tier is composed of 4-5 cells. Hence, the air pore is surrounded by a minimum of 14 and a maximum of 40 cells. Below the upper epidermis, there ts a single row of air chambers. The successive air chambers are separated from "one another" by vertical filaments. Below the air chamber we find several layers of large, thin-walled, parenchymatous cells without chloroplasts. These cells mostly contain starch grains, but there are some mucilage and oil containing cells here and there. From the lower epidermis arise rhizoids and scales (Figure 5.10).

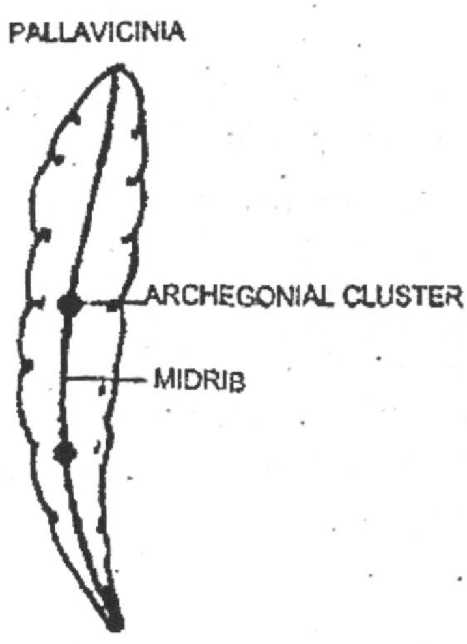

Figure 5.11

Pallavicinia sp.

Thallus has a midrib of many cell thickness; wings are of single cell thickness. No air chamber only smooth walled rhizoids occur (Figure 5.11).

Riccardia sp. (*Aneura*)

The plant body is thalloid dorsiventral and dichotomously branched. The unequal growth of dichotomies resulted the gametophyte in which the branches appear to be lateral appendages. The thallus has insignificant midrib, a few cells in thickness and wings are of single cell in thickness. From the lower epidermis only smooth walled rhizoids arise (Figure 5.12).

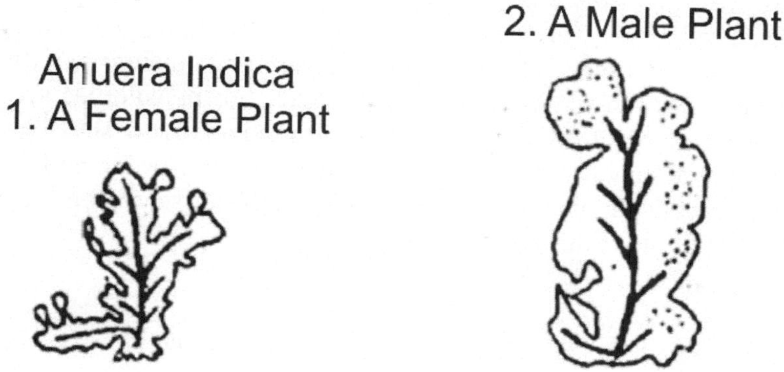

Figure 5.12

Thallus is fleshy, branched, epidermal cells are smaller than the other cells. Sex organs are on short lateral branches. Male branches are more or less circular in outline. Antheridia biseriate. Female branches are very short 2-8 archegonia (Figure 5.13).

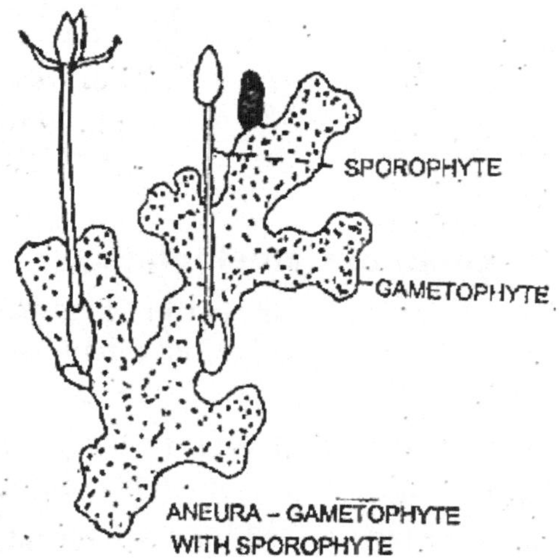

Figure 5.13

Anthoceros

Anthocerotae differ from other Bryophyta in having large chloroplasts, each with pyrenoid; antheridia and archegonia are completely embedded in the gametophyte. Indefinite growth of the sporophyte is due to the presence of meristematic region at the base. There are about 300 species.

The plant body of Anthoceros is dorsiventral, fleshy, dark green, with marginal irregular lobes and without midrib (Figure 5.14). There is no internal differentiation of the thallus. The entire thallus is formed by thin walled parenchymatous cells. From the lower epidermis, only smooth walled rhizoids arise. Scales are altogether absent. Each cell of the gametophyte has a single large chloroplast with pyrenoid as the stored food. There are also some intercellular, bluish-green, dark spots. There are cavities filled with mucilage, in which *Nostoc* colonies are present. These cavities open out on the ventral side through a narrow slit (Figure 5.15). The gametophyte may be. monoecious or' dioecious. Antheridia and archegonia are developed, endogenously and are embedded in the dorsal region of the thallus. Antheridia are found in antheridial chamber.

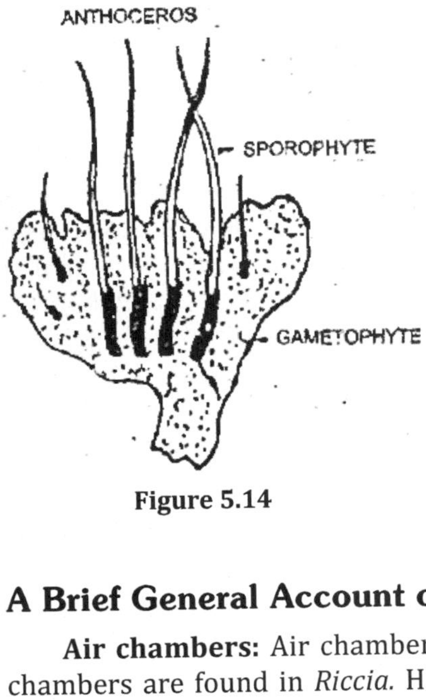

Figure 5.14

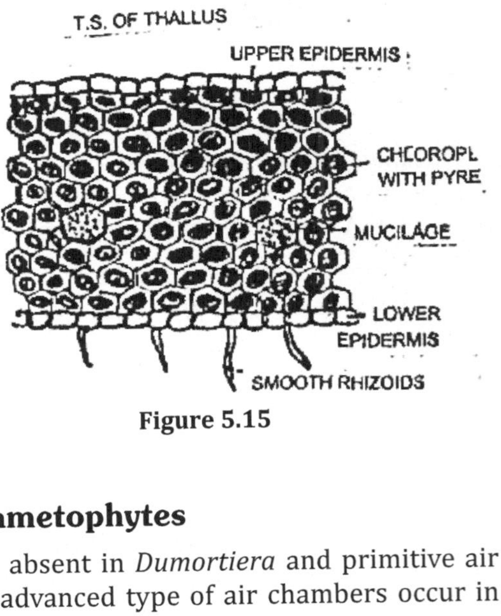

Figure 5.15

A Brief General Account of Gametophytes

Air chambers: Air chambers are absent in *Dumortiera* and primitive air chambers are found in *Riccia.* Highly advanced type of air chambers occur in *Marchantia.*

Air pore: Very primitive type of air pore is found in *Riccia.* Simple air pore occurs in *Reboulia.* Highly complex air pore is found in *Marchantia.*

Assimilatory filaments: Unbranched, long, assimilatory filaments are found in *Riccia.* Branched or unbranched assimilatory filaments of 2-3 cells in height occur in *Marchartia.* Assimilatory filaments are absent in *Reboulia.*

Sex organs: They are sunken in the thallus of *Riccia.* They are elevated on a specialized stalk in *Marchantia.* The archegonical disc is 8-ranked in *Marchantia.* In *Reboulia* male receptacle is sessile.

Gametophytes of Leafy Liverworts-General Account

Majority of members of Jungermanniales belong to this category. The gametophytes of leafy liverworts are differentiated into an axis like, leaf-like and rhizoid-like structures. Since leaf-like structures are present, this group is called leafy liverworts. The gametophytes of Jungermanniales are always foliose and with the leaves borne in a regular spiral succession along the stem. The classification of the Jungermanniales even today is in a very controversial state and there is a great divergence of opinion as to the.limits of the families and the genera included in each family. Based upon the position of archegonia, the -Jungermanniales can be divided into two main groups. They are Acrogynous Jungermanniales (Apical cell being used up in the formation of an archegonia) and Anacrogynous Jungermanniales (apical cells are not involved in the formation of arche. However, this is' an artificial grouping. Growth of the gametophytes liverworts takes place by a single, three-sided apical cell (*Pleurozia* is an exception of this). The apical cell cuts off activates by its three cutting faces, either by clockwise or counter clockwise (anticiock wise sequence). Each derivate contributes to a leaf, a small portion of stem. Hence, the gametophyte has three rows of leaves. A derivative divides by an oblique vertical division, so that a smaller triangular cell and a larger cells are formed. The larger cell again divides transversely so that three cells are formed. The top two cells divide transversely so that five cells ai produced. This is called a merophyte. Of the five cells of merophyte, the top cells contribute to the leaf, the middle two cells form the cortical parts of the and the single lower cell give rise to medulla portion of the stem (Figure 5.16). The gametophyte bears three rows of leaves, of which two are dorsal or lateral. They are present on the dorsal side of the axis or stem. The third leaf occurs on the ventral side of the axis. It is called a ventral leaf or amphigastria. In *Herertia* sp.

In *Porella* sp., the lateral leaves are larger and the ventral leaves are much reduced in size. This is due to the prostrate habit of the plant. The dorsal side, which receives more light, has larger leaves while ventral side which receives only minimum light has smaller leaves. Initially the leaves are inserted transversely to the axis. The ventral leaf always remains transverse but the lateral leaves become diagonal, due to differential growth of the stem. When the growth of the axis is more on the dorsal side the arrangement of leaves is said to be succubous. Here the growth of the axis is said to be hypotropic. In succubous arrangement, the forward edge of one leaf is found below the hind edge of the leaf just above (*Lophocolea*). When the growth of the axis is more on the ventral side, which results in the incubous arrangement of leaves. The growth of the stem is said to be epitropic. In this type, the forward edge of one leaf is resting on the hind edge of the leaf immediately above (*Radula sp*). *In* succubous arrangement of leaves, upward movement of water is easier than in the incubous form. Incubous forms always grow where there is abundant of ground water. Downward movement of water is easier and rapid in incubous form. Hence, these plants always grow on tree trunks as epiphytes, where there, is downward flow of water. Leaves are single cell in thickness. All the cells contain chloroplast. In plants like *Porella sp.,* the dorsal leaves are unequally bilobed, having two lobes. The larger lobe is ovate with almost rounded apex. This is called antical. The smaller lobe is linear and pointed apex. This is called postical. In *Frullania* and *Radula* by the side of the postical lobe there is a smaller outgrowth called stylus. The gametophyte is attached to the substratum by means of smooth walled rhizoids alone, produced from the ventral side of the axis. In *Radula,* the rhizoids are produced from the postical lobes of the lateral leaf. The rhizoids are usually unicellular, rarely multi-cellular as in *Plagiochila*.

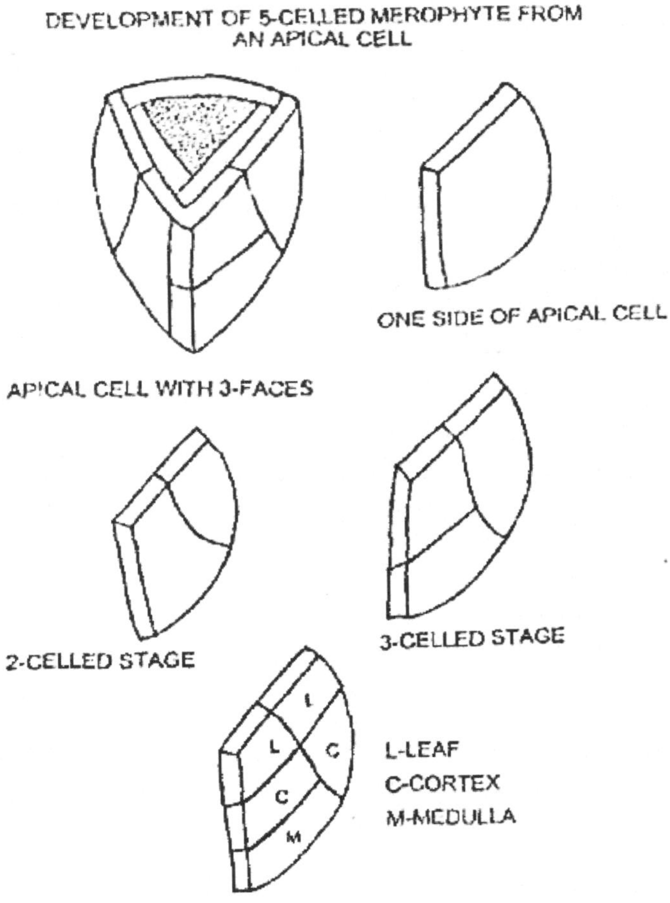

Figure 5.16: Development of 5-celled Gametophyte from an Apical Cell.

Porellaceae-Porella sp

Porella belongs to Porellaceae.

Porella platyphylla is a commonly distributed species. *Porella* grows on moist, cool and shady places. It grows on moist rocks, tree trunks and old walls, and forms a compact greenish patch, practically covering the entire substratum on which it grows.

Structure: It is an acrogynous form. The plant body consists of a slend, dorsiventrally flattened and piustrate main axis with lateral branches. The axis bears three rows of leaves, two rows of dorsal, larger leaves and a row of ventral leaves, which are much reduced in size, called amphigastria. When we observe the axis on the dorsal side, the dorsal leaves completely cover the axis and they are arranged in such a way that the anterior edge of each leaf covers the posterior edge of the leaf next above it. This type of overlapping arrangement of.the leaves is known as incubus arrangement. Each dorsal leaf is unequally bilobed. The larger lobe is usually oval shaped with a rounded apex, is called anticallobe. The smaller lobe is narrow with acute apex is called posticallobe or the lobule. Anatomy of the leaf is very simple in its structure. Each leaf is single cell is thickness without midrib. All the cells have chloroplast (Figure 5.17).

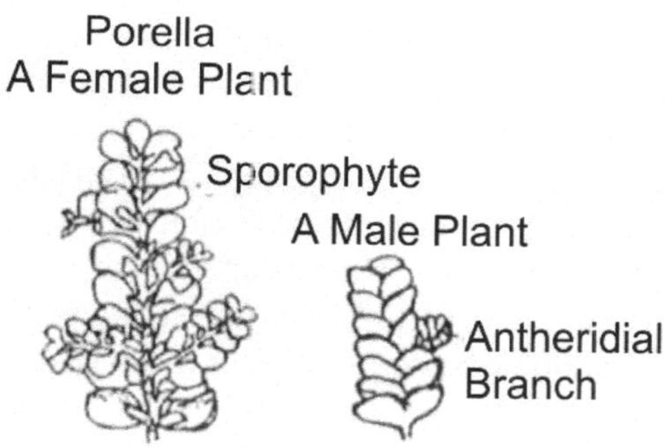

Figure 5.17

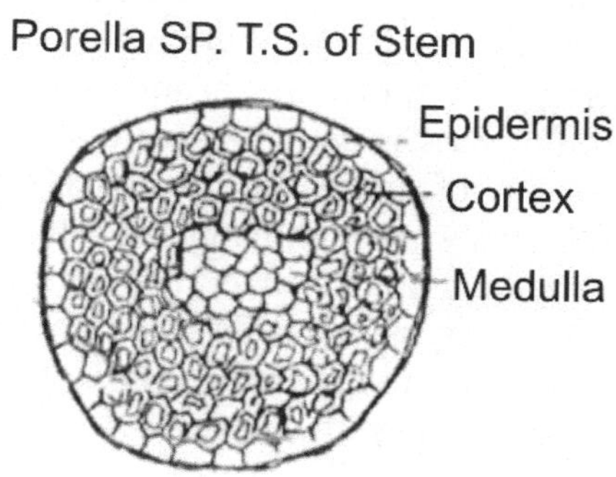

Figure 5.18

T.S. of Axis

Axis is differentiated into three regions, namely epidermis, cortex and medulla, epidermis is single-layered and consists of chloroplasts. Cortical cells are always larger than the medullary cells. Cortex stores food materials. Medulla is for the conduction of water and mineral salts and food materials (Figure 5.18).

Gametophytes of Mosses

Mosses are the higher Bryophytes. Bryopsida is the third and the largest class of Bryophyta. It consists of 660 genera and 14500 species. The Bryophyta are characterised by the following features. The predominent stage in the life cycle is the gametophyte. It is green and independent plant. The gametophyte is differentiated into two stages. A branched prostrate, juvenile filamentous or thalloid stage, the protonema, and an erect, radial leafy shoot, the gametophore, which is usually persistent. The gametophore consists of stem and spirally arranged leaves, and bears sex organs. Protonema die after gametophytes are formed. Rhizoids borne by both protonema and gametophore multicellular and with diagonal cross walls.

Sex organs are unique. Early growth stages are initiated by an apical cell with two cutting faces. Sporophytes have foot, seta and capsul. The wall of the capsule consists of several layers of cells covered by an epidermis. Capsule encloses a central columella. The archesporium gives rise to spores only, elaters and nurse cells are absent. The Bryopsida are usually divided into three orders namely, Sphagnales, Andreaeales and Bryales (Bower, 1935 and Campbell, 1940). Dixon (1932) gave the above orders the rank of Sub-class but retained the sufflx-"ales" in each case.

We shall discuss, selecting a few typical examples for each.

Gametophytes of Mosses

Sphagnales

Sphagnales consist of one family Sphagnaceae which has one genus, *Sphagnum.* There are about 320 species. About 17 species have been reported from India. *Sphagnum* is commonly called *Bog Moss* and *Peat Moss. Sphagnum* is almost cosmopolitan in distribution. It is found in all parts of the world except in Arctic region. The spores on germination will give rise to a simple thalloid protonema. Protonema is irregularly lobed, thalloid, one cell in thickness. It is fixed to substratum by means of multicellular rhizoids (Figure 5.19). Protonema develops into upright, radial, leafy shoot, the gametophyte. The mature gametophyte is variable in colour. It may be orange brown, rose-pink, orange-pink or red. The young gametophyte bears numerous rhizoids, which are multicellular and plate-like and the cross walls are oblique. All species of *Sphagnum* grow in ponds, swamps or other moist places and usually in soft water where there is little lime.

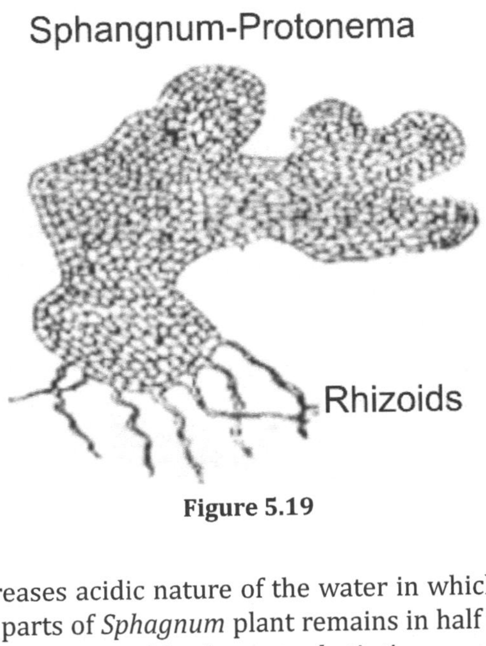

Figure 5.19

Sphagnum increases acidic nature of the water in which it is present. Since the water is acidic, parts of *Sphagnum* plant remains in half rather little decayed condtion. The compacted mass of dead gametophytic tissue accumulatinglas growth continues from year to year, together with the remains of other plants is called peat. Hence, *Sphagnum* is called *Peat Moss.*

The adult gametophore is very slender and weak. The stem has leaves which are distantly located. Tuft of lateral branches arise from every 4th leaf of the main axis. The number of branches varies from 3 to 8 and usually five. The lateral branches are completely limited in growth and are closed and compactly arranged with minute leaves. The leaves of the main axis and the leaves of the lateral branches differ in size, shape and structure (Figure 5.20).

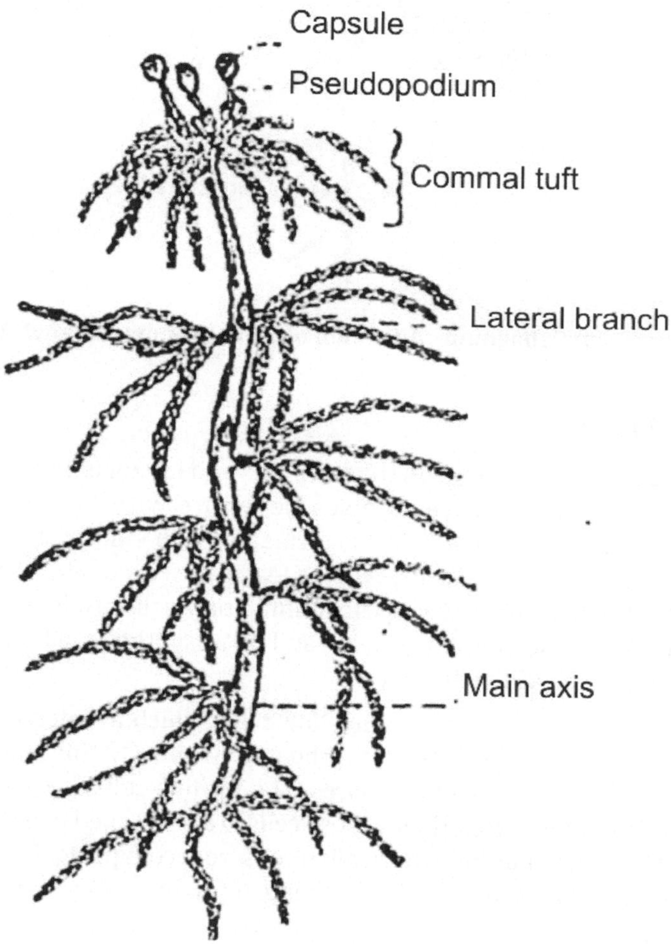

Figure 5.20: Sphagnum Gametophyte.

The lateral branches are of limited growth. Some lateral branches are horizontal to the axis and become arched at the middle with their distal portions curving downward. These branches are called divergent branches. Some lateral branches are drooping down around the main stem. These branches are called drooping branches. These branches act as water conductors (Figure 5.21). The lateral branches at the apex of the stem are stumpy, short and are limited in growth, crowded and clustered into a compact conspicuous head called coma or commal tuft. These are formed by the reduced internodes between the branch tufts, and when axis grows in length, these short branches get elongated and become normal branches.

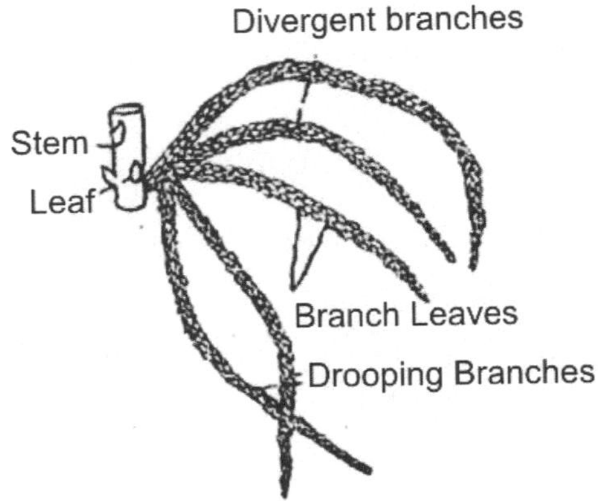

Figure 5.21: Sphagnum - A Portion of Stem to Show a Branch Tuft.

Anatomy of Leaf

The leaves which are found on the main axis and lateral branches are entirely different. In general, the leaves are sessile, entire with smooth margin and acute apex. There is no midrib. A leaf is formed by two kinds of cells namely the assimilatory cells and hyaline cells. When a leaf is observed under microscope, it gives an appearance of a net-work. The assimilatory cells are living, narrow and elongated and have chlorophyll pigments and photosynthetic in function. The net work is formed by the assimilatory cells.

These meshes are occupied by hyaline cells, which are dead cells without protoplast. The hyaline cells are strengthened by spirally thickened bands. On the walls of hyaline cells, there are pores. The hyaline cells are like sponge and they absorb, and retain water. Hence these cells are also called capillary cells. The strengthening spiral bands help the hyaline cells from collapsing. The assimilatory cells and hyaline cells are alternatively arranged (Figure 5.22).

Anatomy of the Axis

The axis (stem) of *Sphagnum* is almost circular in outline. It shows three distinct regions. They are outer cortex, middle hardrome and central axial cylinder medulla. The cortex is single cell in thickness when the axis is young. At maturity it becomes many celled in thickness (Figure 5.23). In *Sphagnum molluscum* and S. *tenellium,*- the cortical cells, on the outside develop some specialized water-storing cells and these cells are called retort cells.

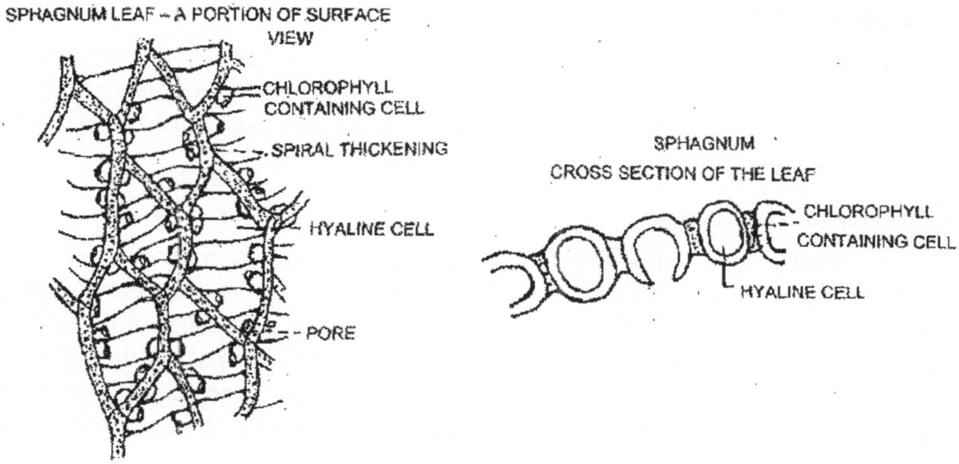

Figure 5.22

Eubrya

These are true Mosses and they differ from other Musci in many respects. In this lesson, we shall study *Bryum* and *Pogonatum* selecting as examples.

Bryum sp.

Bryum is a very common Moss and is distributed throughout the world. *Bryum* belongs to Bryaceae. It is cosmopolitan in distribution, and is the best known of all the Mosses. The gametophyte consists of two distinct phases: 1. a prostrate, filamentous algae-like, ephemeral structure, the protonema and 2. *The* upright, leafy, persistent Moss plant, the gametophore.

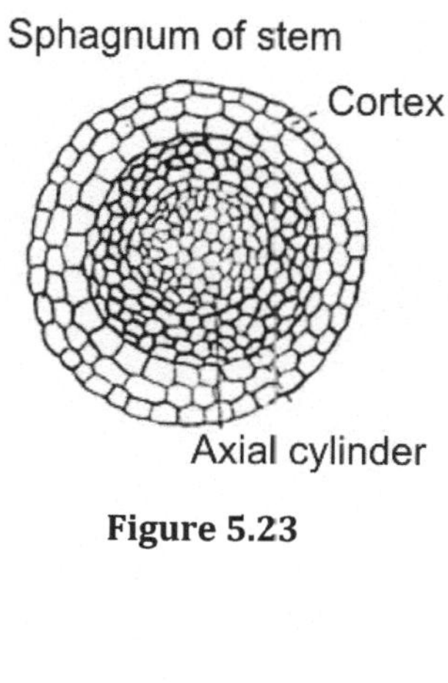

Figure 5.23

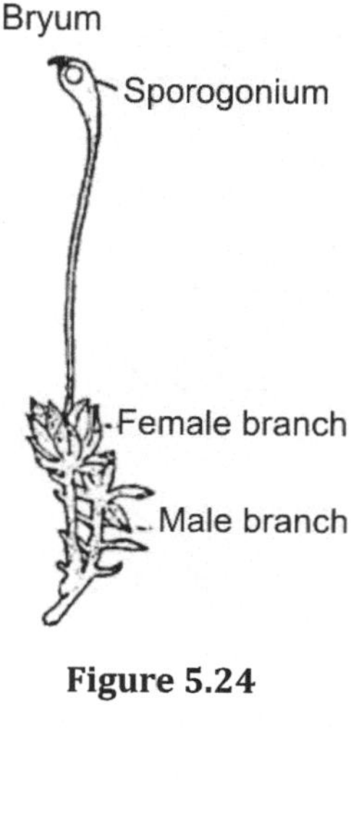

Figure 5.24

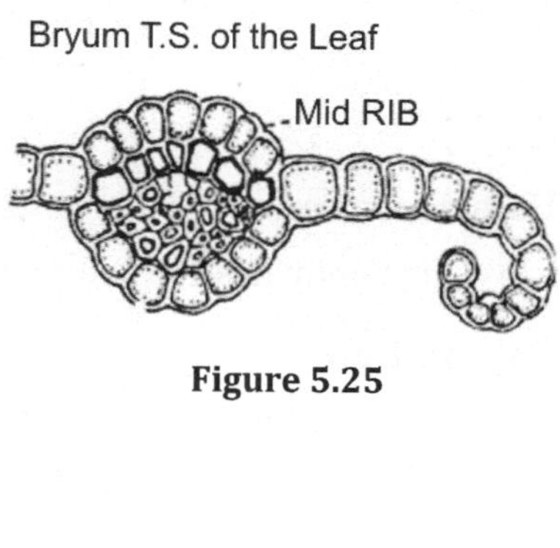

Figure 5.25

Protonema Stage

Spore on germination develops into slender, green, branched, filamentous structure called protonema. It is only vegetative and transitory in this plant. This. has two types of branches, namely, chloronemal branches and rhizoidal branches. Rhizoids are slender branched and multicellular.

Adult Gametophore

The adult gametophore is differentiated into 3 regions namely rhizoids, axis and leaves. Leaves are spirally arranged in 3/8 phyllotaxy. Leaves are entire, sessile, oblong-ovate, broad based, pointed apex and smooth margin (Figure 5.24). The leaf has a well defined midrib which is several cells in thickness, while the wing on either side is single layered. The centre of the midrib is occupied by a small central strand of narrow, thin-walled cells which form a simple type of conducting strand (Figure 5.25).

Stem

The apical growth of-the stem takes place by means of single *apical ceil* with three cutting faces.

A cross section of the stem shows, three well-differentiated regions namely a single layered epidermis, multilayered cortex and central cylinder. The central cylinder consists of long narrow, thin-walled, colourless cells, without protoplasm. The cells of the cortex, in the young condition consist of chloroplasts. In mature stem, the cortex is differentiated into outer cortex with thick-walled cells and inner cortex, with thin-walled parenchymatous cells.

Epidermal cells usually contain chlorophyll (Figure 5.26).

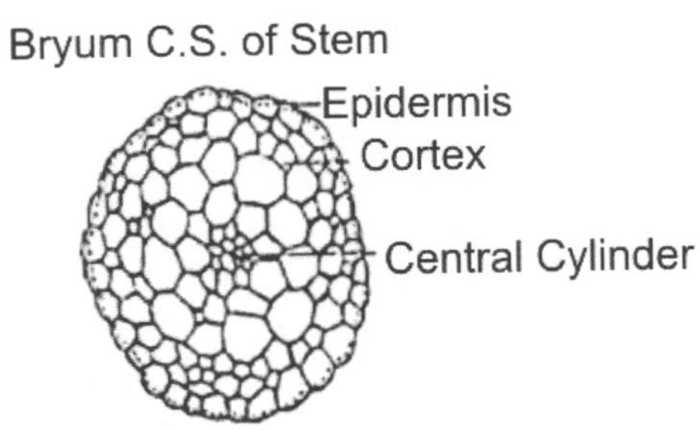

Figure 5.26

Polytrichaceae

This family has 15 genera and 350 species.

Pogonatum

This has about 158 species distributed throughout the world and mainly in cool, temperate and tropical regions. It is cosmopolitan in distribution. *Pogonatum* shows some of the highest stages in the development of the gametophyte and agorophyte generation. The gametophyte is differentiated into rhizome, axis (leafy shoot), leaves and rhizoids. Spores on germination give rise to green, filamentous structure called the protonema. The gametophyte is branched. In the axis, at the base of every 12th leaf there is a lateral bud.

Leaves

The leaves on the rhizome and the middle transitional region are brown or colourless and are arranged in 1/3 phyllotaxy. The leaves on the aerial leafy axis are large, broad, thick, rigid and are arranged in 3/8 phyllotaxy (Figure 5.27).

Anatomy of the Leaf

In a transverse section, the leaf shows the following tissues. The midrib is several cells in thickness at the centre and gradually merges into 6 the rudimentary wing (lamina) at the margins. On the lower surface, there is a well-defined epidermis, made up of large cells whose oilier walls are strongly thickened. Within the epidermis are one or two layers of small, sclerenchymatous, elongated cells, the central tissue of the leaf is formed by large, thin-walled parenchymatous cells. There are some sclerenchymatous cells, here and there. On the upper surface, there is a layer of large chlorophyll containing cells and they are assimilatory in function. Each plate is 5 to 8 cells in height and terminal cell of each plate is wider or papillose and differ from other cells (Figure 5.28).

Pogontum - Habit

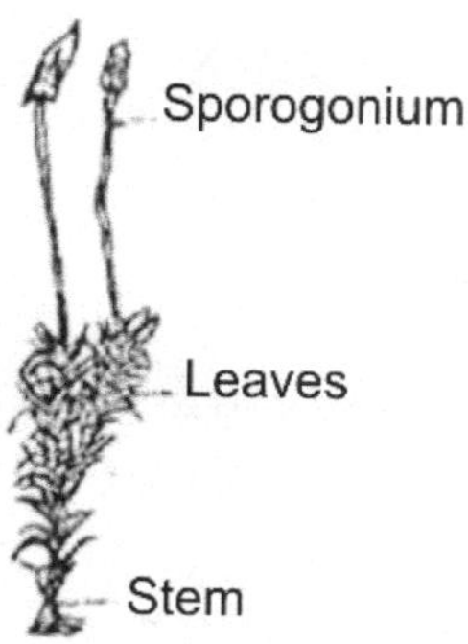

Figure 5.27

Bryum C.S. of Stem

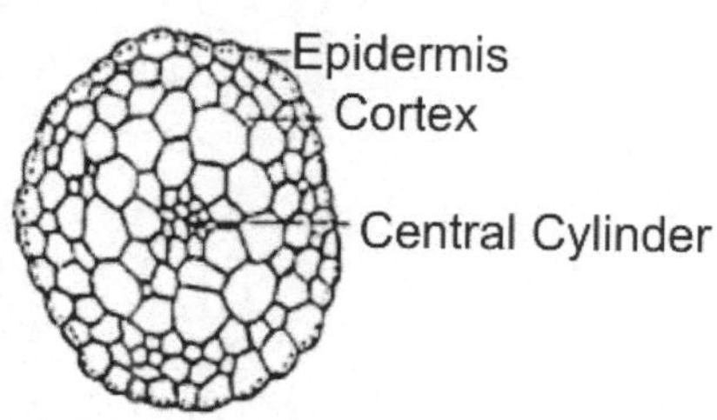

Figure 5.28

Anatomy of Rhizome

Rhizome in outline is differentiated into three regions. They are 1. Piliferous layer 2. Cortex and 3. Central cylinder.

The central cylinder is almost triangular in outline. It is composed of two different types of cells called stereids and hydroids. Around the central cylinder, near the groove region, there is a layer of parenchymatous cells having starch grains. The layer is called amylon layer. Near this layer, there are sieve tube like cells called leptoids. They are surrounded by 2 or 3 layers of pericycle, which is discontinuous. There is a single layer of endodermis which separates the central cylinder from the cortex, which is parenchymatous. Cortex is interrupted by three hypodermal strands which are prosenchymatous cells. They are living cells. There are cells which are thin-walled and lignified. They show no sharp line of demarcation from those of the hypodermal strands and form together with it a wedge-shaped mass of tissue gradually narrowing inwards and are called the radial strands (Figure 5.29).

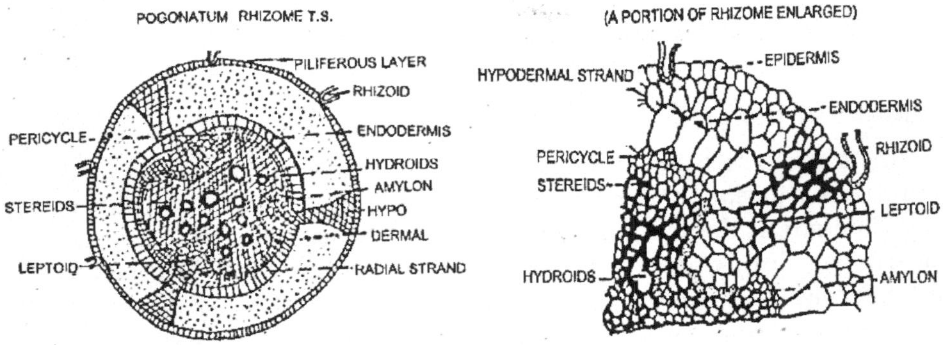

Figure 5.29

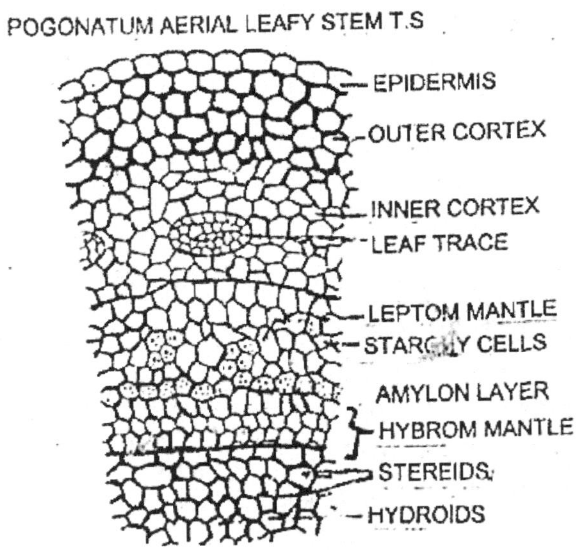

Figure 5.30

Anatomy of aerial leafy stem (axis). The outermost layer is piliferous layer from which numerous rhizoids arise. Axis is more or less'round in outline. Sometimes, it is irregular due to the presence of leaves. The outermost layer is epidermis. This is single cells in thickness. The superficial layer is hard to distinguish. The cortex is differentiated into outer cortex and inner cortex. The outer cortex consists of compactly elongated and prosenchymatous cells. The inner cortex consists of parenchymatous cells. The inner cortex is followed by a rudimentary pericycte. Inside this, there are sieve tube-like cells called leptom mantle. Internal to this amylon layer. Inside the amylon layer there is hydrom mantle. In the centre of the stem is hydrom cylinder which consists of stereids and hydroids and they are concerned with conduction (Figure 5.30).

Reproduction in Bryophytes

i) Asexual and Sexual Reproduction in Bryophytes

a) Asexua/(Vegetative) Method of Reproduction

One of the characteristic features of the Bryophytes is their tendency towards vegetative reproduction by a large variety of means. In many highly specialized forms, especially the dioecious species, the sexual reproduction has been largely abandoned and the propagation of species has become dependent on vegetative reproduction.

Correns (1899) and Cavers (1903) have described the various methods of vegetative reproduction in the Bryophytes, the former in Bryopsida and the latter in the Hepaticopsida. There are different types of vegetative methods of reproduction found in Bryophytes. They are as follows:

1. The progressive growth of anterior portion of the thallus is followed by the death and decay of older, posterior part of the thallus. This is a very common method of vegetative reproduction in thalloid liverworts and Jungermanniales.
2. Formation of adventitious branches from the midrib of the ventral side of the thallus. The branches are detached which produce new thallus. This type of reproduction is found in *Riccia, Metzgeria, Corsinia, Reboulia, Anthocerous* and in many members of Jungermanniales.
3. Formation of innovations and their separation *e.g.* Acrogynous Jungermanniales and *Sphagnum.*
4. Formation of Cladia: There are mainly two types.
   a. Leaf Cladia: These are formed by the individual cells of the leaf *e.g. Plagiochila* and *Frullania.*
   b. Stem Cladia: These are formed from stem *e.g. Leptolejeunea.*
5. Separation of whole plant: It is found in *Pohlia* and seperation of modified branch of bud-like form, as in some species of *Bryum* and *Pohlia.*
6. Formation of tubers is found in *Riccia, Anthoceros, Fossombronia* and *Pallavicinia etc.*
7. Formation of Gemmae: Gemmae are specialised structures meant for vegetative propagation, produced in a specialised cup-like structures

called gemma cups. Gemma cups are discoid in *Marchantia,* (Figure 5.31) semilunar in *Lunularia* (Figure 5.36). Spherical gemmae are produced in flaskshaped receptable in *Blasia* (Figure 5.34), discoid and multicellular gemma in *Radula and Metzgeria* (Figure 5.35) one to three celled in *Lophozia, Lophocolea,and Riccardia* (Figure 5.32) multicellular-articulalte in *Ulota,* (Figure 5.33) globular, multicellular in *Bryum.* Solitary gemma may be formed on rhizoids, either those of protonema or those of the leafshoot. Subterranean gemma are usually called bulbils.

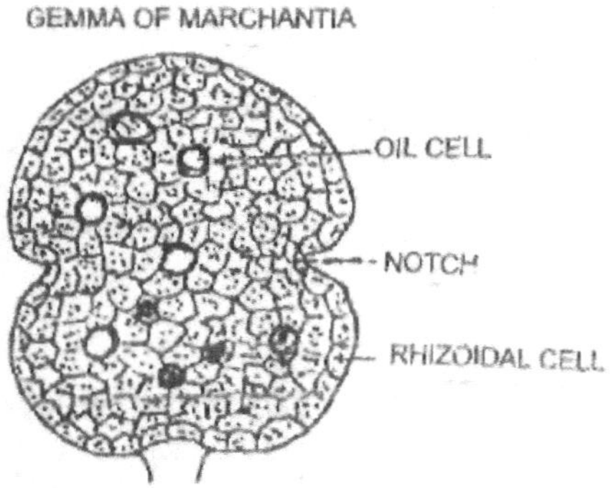

Figure 5.31

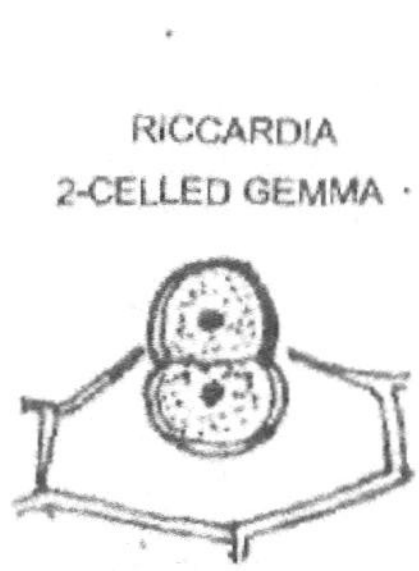

Figure 5.32

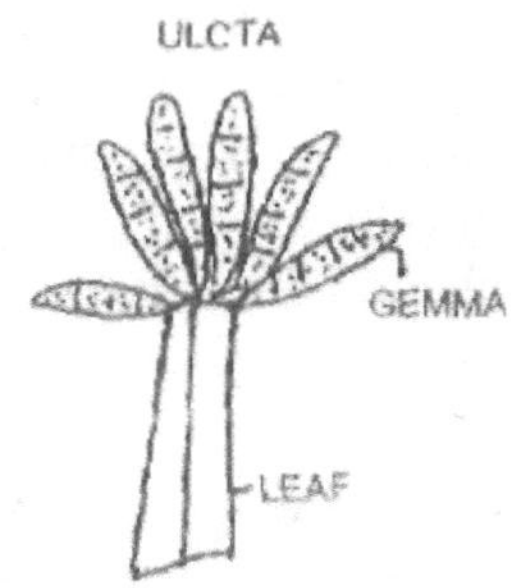

Figure 5.32

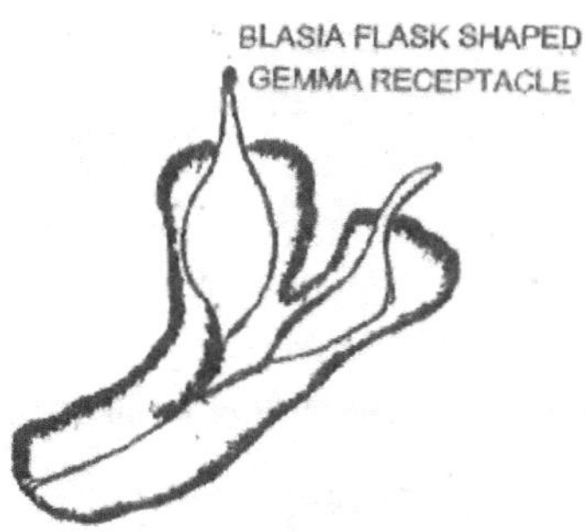

Figure 5.34

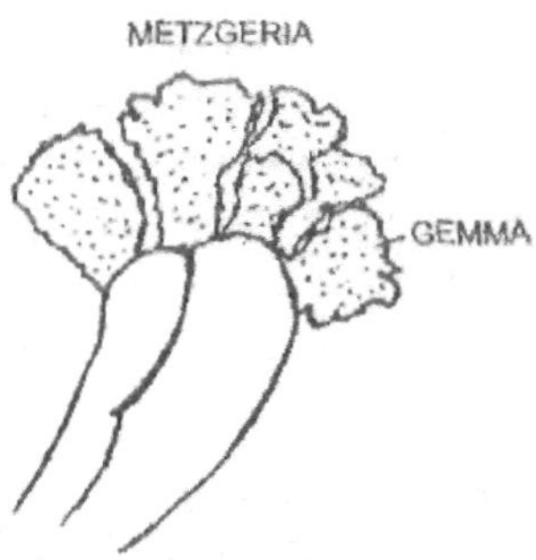

Figure 5.35

8. Formation of Primary Protonema in Bryideae: A protonema developing from a spore usually forms several buds. Development of the buds into leafy shoots, followed by death of the protonema, results in several independent gametophores. The number of gametophores may be increased by multiplication of the protonemal stage. Death of cells here and there in a protonema results in short fragments, each of which may grow into a new protonema.
9. Formation of Secondary Protonema in *Sphagnum:* When growing in light, rhizoids of a leafy shoot may become protonemal in nature and produce buds just as do protonemata from germinating spores. Development

of the buds into gametophores followed by death of the protonema like rhizoids, results in new independent gametophores. Secondary protonema may also be developed from wounded portions of the leafy shoot. The wounded tissue may be that of a leaf or stem.

10. *In Leucobryum glaucum,* occupying the centre of a terminal rosette of leaves in a tuft of rhizoids and from the rhizoids, new plants arise which subsequently fall off and give rise to new plant.

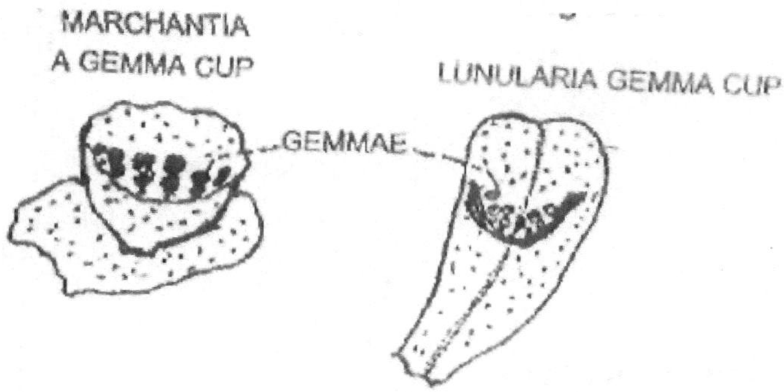

Figure 5.36

11. Apospory: Wounding of sporophytic tissue may result in the production of protonema. The gametophytes thus developed are aposporous in origin, since they are formed from the sporophyte without spore formation. Notwithstanding the fact that their cells have the diploid number of chromosomes, the aposporously produced protonema and gametophores are normal in appearance and with fertile gametes. Fertilization may take place between gametes of two diploid gametophores, or between gametes from a diploid and haploid gametophores. According to the origin of the gametes, the new sporophyte has 4n or 3n chromosomes.

b) Sexual Reproduction in Bryophytes

The sex organs are antheridia and archegonia.

The Antheridia: Antheridium is the male sex organ. The position and protection of antheridia varies in different species.

In *Anthoceros* and *Notothylas,* antheridia are endogenous and are embedded in the dorsal surface of the thallus in closed cavities called antheridial chamber, the roof of the chamber is 2 celled in thickness. More than one antheridia are found in each cavity. In *Riccia,* they are sunken in the gametophytic tissue. In *Riccardia,* antheridia are borne on special short lateral branches. In *Pallavicinia,* antheridia are borne in two rows, one on either side of the midrib on the upper surface of the thallus. Each antheridium or a group of two or three is substended by a scale arising from the midrib.

In *Sphaerocarpos* and *Geothallus,* antheridium is protected by a sessile, flask-shaped involucre. The wall of the involucre is made up of a single layer of cells. In *Riella,* antheridia are found along the margin of the wing and they are protected by an involucre.

In *Plagiochasma,* there are many antheridial groups (sessile receptacles) on the dorsal surface of the thallus and they are protected by linear scales.

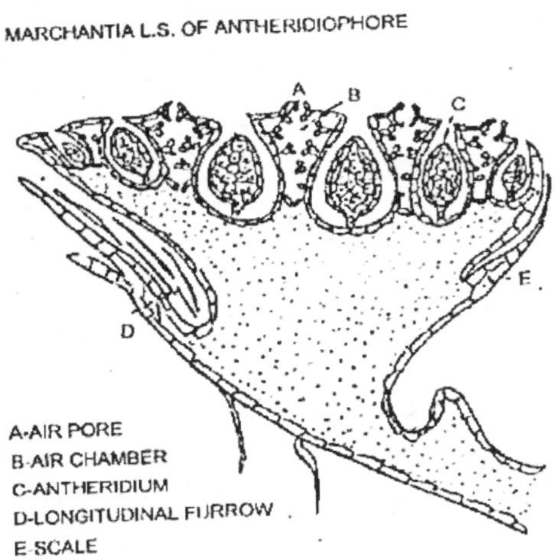

Figure 5.37

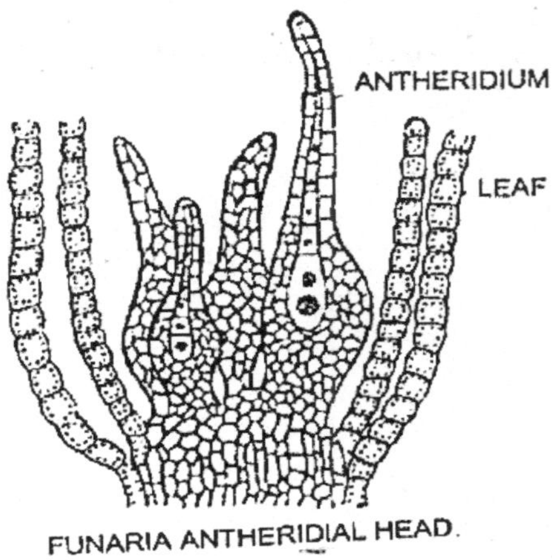

Figure 5.38

In *Targionia,* antheridia occur on specialized, short, adventitious, ventral branches. In *Marchantia,* antheridia are elevated on a specialised branch called antheridial disc (Figure 5.37). In *Dumortiera,* they are sessile. In acrogynous Jungermanniales and in Sphagnales, antheridia arise in the axils of modified leaves called perigonial bracts. In most of the Mosses, antheridia are found at the top of the branch called antheridial branch. In Mosses, antheridia are *very* long and elongated with a *very* long stalk. Antheridia are surrounded by paraphyses.,. Paraphyses are multicellular, sterile, hair like structures (Figure 5.38). The tip portion is globose or swollen. The paraphyses and antheridia together give an appearance of a false flower. So, it is called 'Moss flower'.

Structure of Antheridium

The size and shape of the body of antheridium varies from plant to plant. Antheridium has a short multicellular stalk and a spear-shaped body. The body. consists of a jacket layer which is single cell in thickness in most Bryophytes except in *Porella* where it is two or three layered at the base. The body encloses a central mass of androcytes and each androcyte mother cell will divide diagonally to form two triangular cells. Each one of this triangular cells will metamorphose into a biflagellate antherozoid.

The Archegonia

Archegonium is the female sex organ. In *Sphaerocarpos* and *Geothallus,* archegonia are covered by involucre. In *Riella,* archegonia are found near the axis. In *Riccia* they are sunken in the gematophytic tissue. In *Pellia,* they are borne on the dorsal side of the thallus. In *Targionia,* they occur on specialised branches. In *Marchantia,* archegonia are held in a specialised branch and are elevated on that branch called archegonial disc (Figure 5.38). In *Anthoceros,* archegonia are completely embedded in the gametophytic tissue (Figure 5.39). In Mosses, archegonia occur on the archegonial head. Among Mosses, the plants may be either monoecious or dioecious (Figure 5.41).

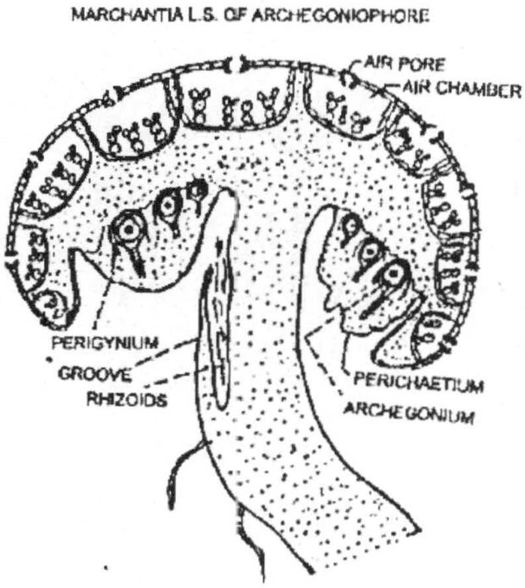

Figure 5.39

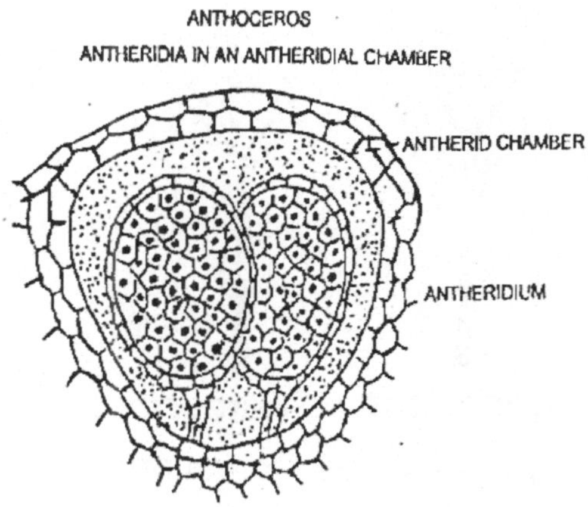

Figure 5.40

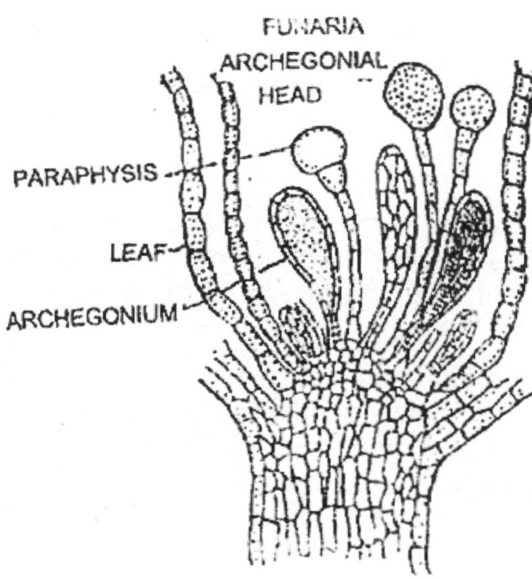

Figure 5.41

Monoecious

The male and female sex organs are found in the same plant. There are three types of arrangements in this, namely, Autoecious, Paroecious and Synoecious conditions (Figure 5.42).

1. The male and female sex organs may be found on two different branches of the same plant (Autoecious condition)
2. Both the sex organs may be found on the same branch and they are separated by one or two leaves (Paroecious condition)
3. Both the sex organs may be found intermingled on the same head (Synoecious condition)

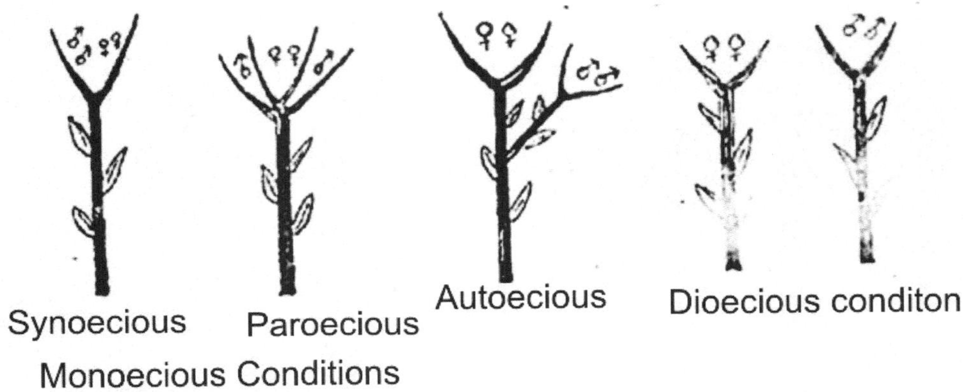

Figure 5.42: Diagram Showing the Arrangement of Sex Organs in Mosses.

Dioecious

The sex organs occur on different plants. Male and female gametophytes are recognised.

Structure

Archegonium has a swollen venter and a narrow neck. The cells forming neck are called neck cells. Neck encloses inside it the neck canal cells. Within the venter, a venter canal cell and an egg cell are present. The neck at the top is well protected by four cover cells.

The neck is usually short and narrow and sharply differentiated from the venter in many Hepaticopsida and Anthocerotopsida. In Bryopsida, the neck is long. The single layered jacket of the neck is composed of 4 vertical rows of cells as in Calobryales and in *Takakia,* five vertical rows of cells are in ungermanniales and six vertical rows of cells as in Marchantiales, Sphaerocarpales and Bryopsida.

The number of neck canal cells varies as follows:

Four in Riccia and Sphaerocarpos, 4 to 6 *Riccardia,* 6 to 8 in *Fossombronia,* 8 to 16 in *Treubia.* Many in Bryopsida.

Evolution of Sporophyte in Bryophytes

The sporophyte of Bryophytes at first is simple in organisation and is attached to the gametophytic tissue. It depends on the gametophyte for its nutrition either totally or partially. Its main function is production and dispersal of spores. Gradually complexity in the structure of sporophytes resulted to perform different functions in addition to earlier mentioned functions. The process of sterilization of sporogenous tissue is very conspicuous in Bryophytes. According to the theory of sterilization. Bower states that the more complex sporophytes have evolved from the simple one by progressive sterilization of potentially sporogenous tissue.

The evolution may be traced from the simplest sporophyte found in *Riccia* to the more complex sporophyte found in Mosses (*Pogonatum*). There is never increasing sterilization of sporogenous tissue. Some of the sporogenous tissue instead of forming spores they remain sterile. The so formed sterile tissues function in different ways, such as foot may perform the function of anchorage and absorption of food materials, chlorophyllous tissue, stomata and intercellular spaced are meant for photosynthesis, elaters, operculum and peristomial teeth in the dehiscence and effective dispersal of spores, the columella in advanced Mosses is used for storage of water and food and in *Anthoceros* the columella is mechanical in function and helps in the dispersal of spores. The progressive sterilization of potential sporogenous tissue resulted in the elaboration and evolution of sporophytes in Bryophytes. Cavers, Campbell and Smith supported the "Theory of Sterilization" proposed by Bower.

Sporphytes of Thalloid Liverworts

The sporophyte in general, is composed of three distinct regions, namely, the foot, seta and capsule. Each part of the sporophyte has its own function.

Foot

It attaches the sporophyte with the gametophyte. It also absorbs food materials from the gametophyte and supplies them to the growing sporophyte.

Seta

It is useful to bring the capsule out of all the coverings, so that the capsule gets exposed.

Capsule

It is the proper spore producing part of the sporophyte. In consists of jacket, spore and elaters and in some cases there is columella. Male and female gametes fuse to form zygote. Zygote is diploid. Zygote is the first cell of the sporophyte. Very simple and primitive type of sporophyte is found in *Riccia* and highly advanced and complex type of sporophyte is found in *Anthoceros.* At the time of spore formation, meiosis takes place. We shall study here some important types of sporophytes.

1. Sporophyte of *Riccia*

In *Riccia,* the sporophyte is very simple and it is only a spore sac. The wall of the archegonium becomes wall of the sporophyte. The spores are covered by the venter of the archegonium. The sporophyte lacks its own covering. The amphithecium which normally is to produce the jackets, gets disintegrated. The sporophyte is undifferentiated into foot, seta and capsule, hence it is very simple in its organisation. The sporophyte is totally parasitic upon the gametophyte for nutrition and shelter. The spores are dark in colour at maturity. Each spore has an exosporium, mesosporium and endosporium. The sporophyte is as a whole simple and primitive. Not at all the sporogenous cells form the spores. Majority of the sporogenous cells form the spores and a few cells form nurse cells. The nurse cells are considered to be the fore-runners of elaters. The sporophyte lacks a dehiscence mechanism. But only by the disintegration of the thallus, the spores are dispersed (Figure 5.43).

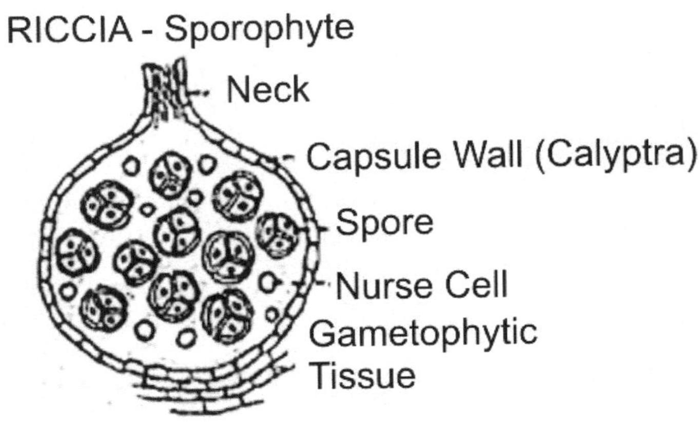

Figure 5.43

2. Sporophyte of *Targicnia*

The sporophyte is differentiated into a bulbous foot, a narrow seta and a large capsule. Jacket layer of the capsule is single cell in thickness. Capsule encloses Spores and elaters, which are formed from the endothecium (Figure 5.44).

3. Sporophyte of *Dumortiera*

The sporophyte of *Dumortiera is* similar to *Reboulia* in shape and structure. It has a well developed foot, seta and capsule Foot is well embedded in the gametophytic tissue. The jacket of cupsule is single cell in thickness and it has spores and elaters in it. The capsule breaks through the calyptra. The capsule wall breaks open and the spores and elaters are exposed and spores are effectively dispersed.

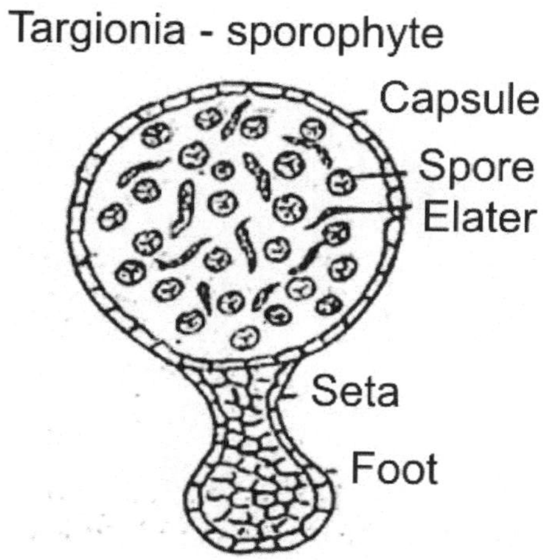

Figure 5.44

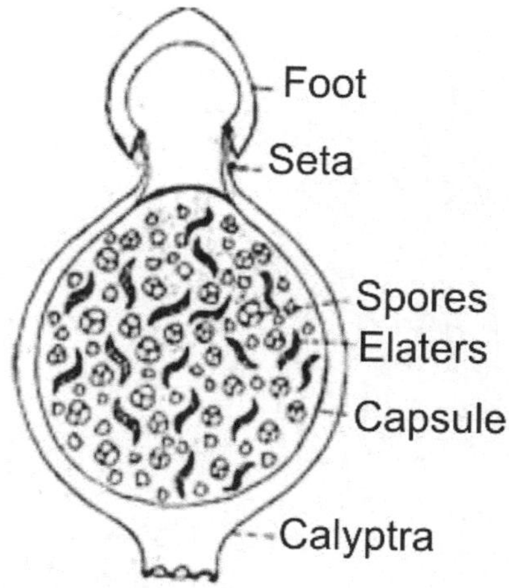

Figure 5.45

4. Sporophyte of *Reboulia*

The sporophyte of *Reboulia* (Figure 5.45) is similar to that of *Marchantia.* It has a well developed foot, seta and capsule. Foot is well burried in the gametophytic tissue. The jacket layer of capsule is single cell in thickness. Inside the capsule, spores and elaters are present. The sporophyte is well protected by calyptra. The spore out put is very less when compared with *Marchantia* and it.as been estimated as 2.500 spores in *Reboulia* and 3,00,000 in *Marchantia*

5. Sporophyte of *Marchantia*

Sporophyte of *Marchantia* is highly specialised. The sporophyte is differentiated into a very large bulbous foot, long and elongated seta and an oval shaped capsule. The jacket layer of the capsule is single layered. Some of the cells of the jacket contain chloroplast. Hence, the sporophyte can prepare atleast "minimum of food materials required by it. Within the capsule, spores and elaters are present. They are formed from the endothecial layer. The sporophyte has its own covering. The sporophyte is adequately protected by calyptra and perigynium and these protective layers are formed from gametophyte. A group of sporophytes are well protected and covered by' a double lipped pendent, fringed involucral sheath called perichaetium, Dehiscence of sporophyte starts from the basal region and extends upto the middle region.

The spores are dispersed out by the hygroscopic movement of elaters. Hence the sporophyte of *Marchantia* has a special dehiscence mechanism (Figure 5.46) (the capsule dehisces irregularly from the apex to about the middle into a number of segments and spores are discharged. Under humid conditions elaters adopt a twisting movement and push the spores out of the capsule).

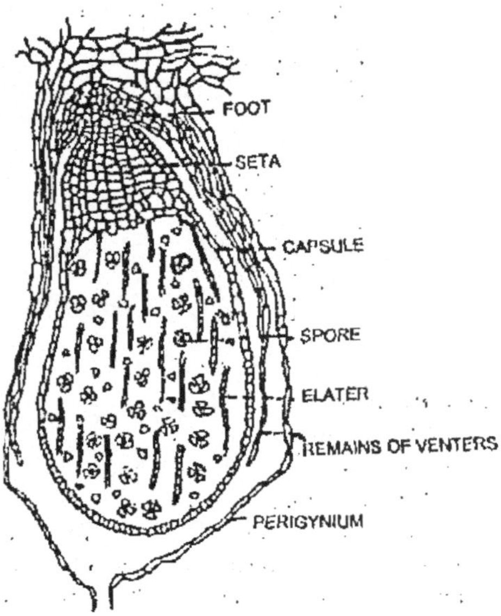

Figure 5.46

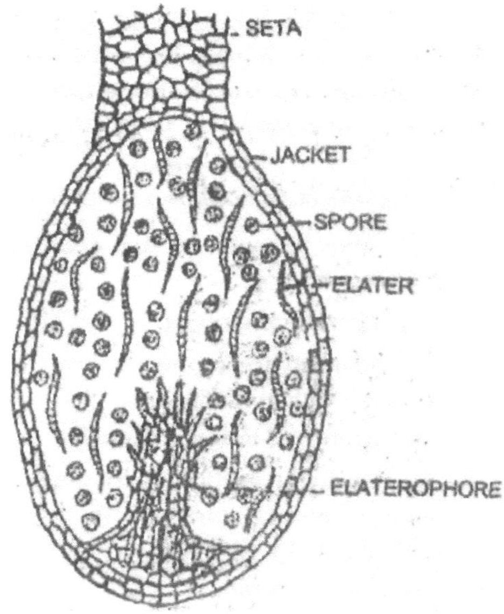

Figure 5.47

6. Sporophyte of *Pallavicinia*

The sporophyte of *Pallavicinia* is almost cylindrical in shape. It has foot, seta and capsule. Foot is indistinguishable. The capsule remains surrounded by a jacket layer which is two cells in thickness. Enclosed by the jacket layer are the spores and elaters.

7. Sprophyte of *Riccardia*

The sporophyte of *Riccardia* has foot, seta and capsule. Jacket layer of capsule is multilayered. Capsule has free elaters and spores. Group of elaters (elaterophore) are present at the tip of the capsule. Spores and elaters are formed from endothecium. During dehiscence the capsule splits into four valves. The elaters occur at the tip of each valve (Figure 5.47).

8. Sporophyte of *Anthoceros*

Sporophyte of *Anthocerous* is highly advanced and specialized. The sporophyte is differentiated into a massive foot, a meristematic region, and a capsule. Due to the presence of intercalary meristem, the growth of the sporophyte is continuous and indefinite. The capsule is differentiated into three regions namely the jacket, sporogenous layer and the columella. The endothecium totally contributes to the formation of columella and part of capsule. The amphithecium by periclinal division forms an outer jacket and inner sporogenous layer. The jacket is multilayered. The outermost layer has stomata. Inner layers have chloroplasts. Hence, the sporophyte can prepare a portion of its required food material. It has reached a sort of physiological independency. At the very base, there is a single layered sporogenous cells. Then it becomes two or three-layered. Still higher up, a difference between the spore mother cells and pseudoelater mother cells, are seen and finally spores are formed from the spore mother cells. The pseudoelaters are branched and septate structures (Figure 5.48).

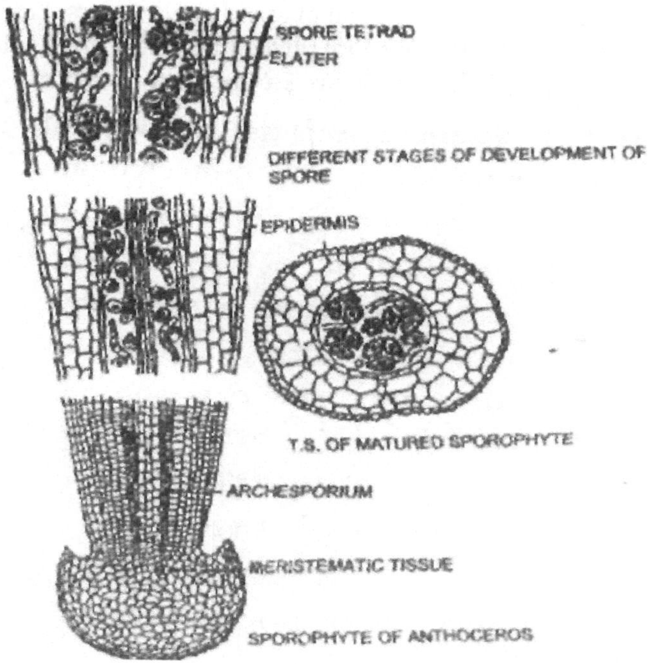

Figure 5.48

In Marchantiales, the elaters are single celled and show bispiral thickening. They are meant for the dispersal of spores only. They perform dual functions. Initially, they help and in the nutrition of spores and at maturity, they help in the dispersal of spores. The development of the sporophyte is successive and not simultaneous.

The sporophyte of *Anthoceros* is advanced on the following grounds.

1. Indefinite growth of the sporophyte due to the presence of intercalary meristem.
2. Spore producing region is very much reduced.
3. The sporophyte prepares its own food by photosynthesis. Thus, it has attained a physiological independency.
4. Occurrence of broad columella.
5. By the disintegration of the gametophytic thallus, the foot of the sporophyte comes indirect contact with the substratum. Now, the sporophyte leads an independent existence.

9. Sporophyto of *Porella*

Sporophyte of *Porella* has a foot, which is merely an enlarged base of the seta. It is indistinct, subtending a short seta and almost a spherical capsule. The sporophyte is well-protected by calyptra, perianth and involucre. Jacket layer of the capsule is many celled in thickness. The entire endothecial cells form sporogenous tissue. They contribute to the formation of spores and elaters. Elaters are short with tapering ends with one to two spiral thickening. There are no elaterophores; capsule breaks open by four valves (Figure 5.49).

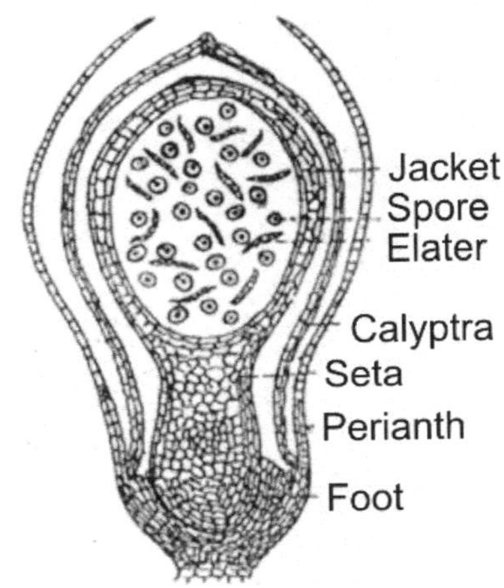

Figure 5.49

Division of the zygote is transverse. The genus *Porella,* has the hypobasal cell developing into a small haustorium of two or more cells, and has the mature sporophyte developing entirely from the epibasal cell. In other genera both the epibasal and hypobasal cells contribute to the mature sporophyte.

Sporophytes of Mosses

The sporophytes of Mosses are generally differentiated into three regions namely, the foot, seta and capsule. The foot is for fixation and absorption of water and mineral salts from the gametophytes. The seta is very long and slender. It elevates the capsule. The capsule is the proper spore producing region of the sporophyte. The wall of the capsule consists of several layers of cells covered by an epidermis with functional or non-functional stomata. There is a central columella. The archesporium gives rise to spores only and the elaters are absent.

1. Sporophyte of *Sphagnum*

Sporophyte is found on a gametophytic stalk called pseudopodium. It hs many perichaetial leaves. The pseudopodium elevates the mature sporogonium *i* its top, beyond the perichaetial leaves much in the same way as does the seta i many other Mosses. The sporophyte is differentiated into two regions namely, th foot and capsule. The foot becomes large and bulbous. Fixation and absorption at the main functions. The seta is absent, and it is represented by a short neck like connections. It is commonly called the neck. The sporophyte is completely covere by the calyptra. The capsule is large and bulbous. It is dark brown or black in colour. There is a central columella which is over-arched by a dome shaped spoi sac, containing spores. The sporogenous layer is derived from amphitheciun Columella is derived from endothecium. The jacket of the capsule is many cell is thickness and the outermost layer is epidermis. The cells of the wall may hav chloroplasts. At the top of the capsule, there is a circular convex disc-shape operculum, which is separated from the rest of the capsule by a circular groove with thin walls, the annulus (Figure 5.50).

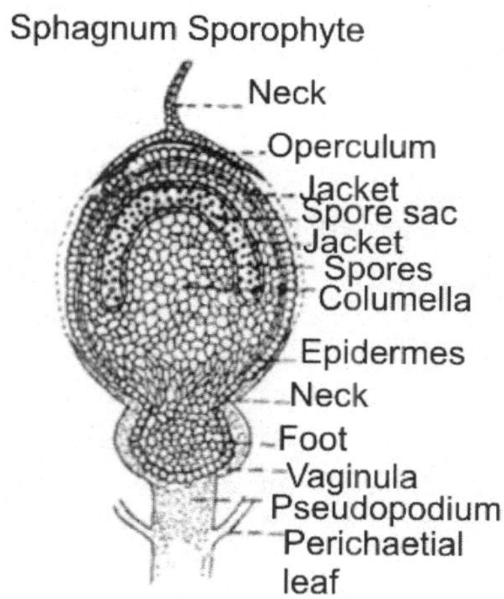

Figure 5.50: Sporophyte of *Sphagnum*.

When the sporophyte matures, it dehisces at the annulus. Operculum is separated from the capsule and falls as a lid. The spores, which were kept under pressure, get liberated through the outcoming air. The capsule bursts open and the spores are forcibly ejected out in the same way as a bullet from an air gun. This method of spore discharge is called air-gun mechanism. There are no peristomial teeth and elaters.

2. Sporophyte of *Bryum*

The sporophyte of *Bryum* consists of foot, seta and capsule. Foot is poorly developed, small, dagger-like, conical structure embedded in the apex of the archegonial branch, from where is absorbs water and mineral nutrients for the developing sporogonium. The seta is a long, slender stalk which bears the capsule.

The capsule is a complex body, and more or less pear shaped. The longitudinal section of the capsule shows the following regions.

Operculum

This is a lid-like structure of the capsule and it lies on the top of it. It is composed of a few layers of cells in thickness. When the capsule dehisces, the operculum is separated as a circular, cup-like lid.

Annulus

This is a special ring-like layer of epidermal cells, lying around the capsule, at the base of the operculum. It is by the rupture of the annulus that the capsule dehisces.

Peristomial Teeth

When the operculum separates and falls out, the top portion of the capsule is exposed. There are two rows of tooth-like projections, ornamented with thick transverse bars called peristomial teeth. The peristomial teeth are hygroscopic and, when they are dry, they open out and help in the dispersal of spores.

Columella

The central portion of the capsule is the columella. The upper end of the columella is cone-shaped.

Spore-sac

Columella is surrounded on either side by a barrel-shaped spore-sac. The spore-sac has an outer wall of 3 to 4 layers of cells without chloroplasts and an inner wall of a layer of cells. Between these two wall layers, the spore mother cells are found.

Air Cavity

Surrounding the spore-sac, there is a wide-cylindrical air space and it is transversed by delicate strands of cells called the trabeculae.

Capsule Wall

Just outside the air-cavity, the capsule is surrounded by 2 to 3 layers of cells. There is a well-defined epidermis. The two outer layers of wall, just below the epidermis, consist of a compact, colourless parenchymatous cells forming a hypodermis, while the innermost layer is composed of loosely arranged cells containing chloroplasts. The colourless hypodermal tissue of the capsule wall thins out below the fertile region, whereas the green inner tissue increases in thickness and is continuous with the green tissue of apophysis.

Apophysis

The seta expands above to form a solid basal portion of the capsule. This is the assimilatory zone. In the centre of the apophysis there is a conducting strand. There is a distinct epidermis having a few true stomata, which are structurally similar to those of higher plants. They are protected by two guard cells. Surrounding the central cylinder and below the epidermis, there is a broad, spongy, green tissue, with intercellular spaces. These cells have abundance of chloroplasts (Figure 5.51). At maturity, operculum is removed as a lid. By hygroscopic movements, the spores are liberated. Peristomial teeth play a major role in the dispersal of spores.

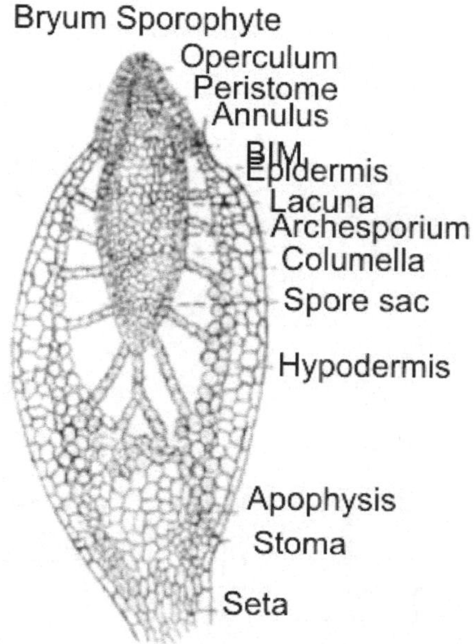

Figure 5.51

3. Sporophyte of *Pogonatum*

Mature sporophyte consists of a foot, seta and capsule. The foot is embedded in the gametophytic tissue. Just above the foot and continuous with it, is the long, slender seta, which supports the capsule at its apex. At the base of the capsule, the seta enlarges to form the apophysis. Apophysis has a distinct epidermis with stomata. The number of guard cells is usually two.

The wall of the capsule is composed of several layers of chlorophyllous cells. The outermost layer is the thick walled epidermis. There is airspace transversed by filaments, which are connected internally with the outer wall of the spore-sac. The spore-sac is limited internally by air space. There is a central columella.

The spore-sac is bounded by thin-walled cells on each side. All the cells of sporogenous tissue give rise to spores. At the top of the capsule there is an operculum. A distinct annulus is not found, but thick-walled rim is present. This is the diaphragm. At the base of the operculum, there is a transverse band of thin membranous tissue composed of cells without intercellular spaces. This tissue stretches like a drum-head over the opening of the capsule and closes it. This is the epiphragm. At maturity the peristome is composed of 32 to 64 short peristomial teeth. Operculum is removed and the spores are liberated. The peristomial teeth are useful in the dispersal of the spores (Figure 5.52). Based on the presence or absence of peristome, the sporophyte of Mosses are classified as Cleistocarpous Mosses or

Stegocarpous Mosses. In Cleistocarpous Mosses, the seta is short, the capsule lacks operculum, peristome and annulus. Capsule opens by irregular splitting and not by operculum or any other regular method (*e.g., Archedium and Acaulan*).

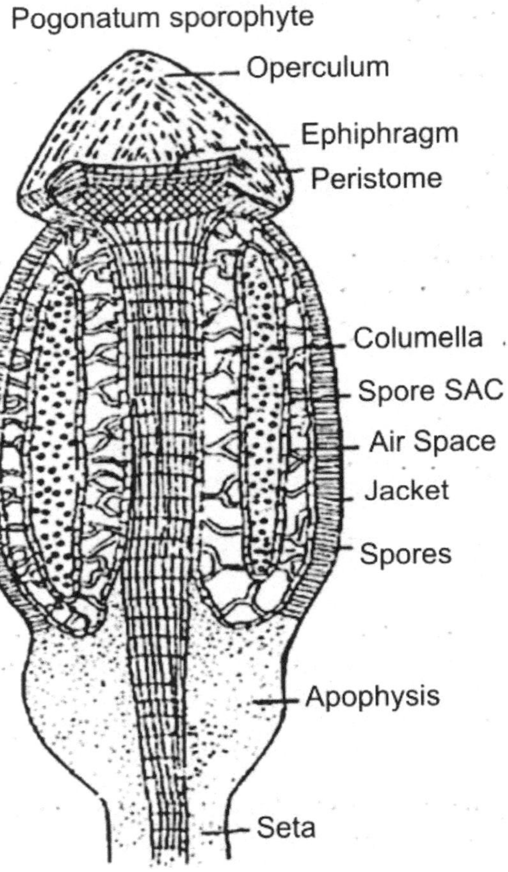

Figure 5.52

In Stegocarpous Mosses, operculum and peristome are present. The Stegocarpous Mosses are further classifiec' into Mosses with Nematodontous peristome and Mosses with Arthodontous peristome. In the Nematodontous type, the peristome-forming zone consists of several concentric layers of cells of the amphithecium. The peristomial teeth are solid, consisting of bundles of whole dead cells and are without any external markings (*e.g. Polytrichum*). On the other hand, in the arthodontous type, the peristome-forming zone consists either of two or three layers of cells of amphithecium. The peristomial teeth are thin, membraneous, articulated and with markings. The peristome may have either one or two rows of teeth. The arthodontous peristome may be further divided into Haplolepidous and Diplolepidous peristome. In haplolepidous type, the peristome consists of a single row of teeth (*Dicranum*). In diplolepidous type, the peristome has two rows of teeth. The sporophytes of Mosses are also grouped into primitive, intermediate and advanced one.

In the primitive type, there is no peristome. In the intermediate type there is a single row of peristomial teeth. In the advanced type of Mosses, the peristomial teeth are in two rows.

Fossil Bryophytes

As there is no lignified vascular tissue or cutinized epidermis in the thallus and with relatively fragile nature of the plant body, the Bryophytes were not well suited for fossilization and hence only scanty or meagre fossil Bryophytes are available.

Fossil Hepaticopsida

The existence of some fossil Bryophytes from Upper Carboniferous period was reported for the first time by Walton (1925). He described five species *of Hepatieites.* They are *Hepatieites kidstoni,* H. *willsi, H. langi,* H. *lobatus,* and *H. metzgerioides. Hepatieites kidstoni* resembles the living genus *Treubia* in respect of the arrangement of leaves. *H. willsi* and *H. langi* had thallus organisation similar to, that of living *Riccardia. H. metzgeria* closely resembles the present day *Metzgeria.* During Devonian period, the fossil Bryophytes were very rare, but, *H. devonicus* has been reported from Upper Devonian which resmbles *Pallavicinia.*

In Mesozoic era *Hepaticites laeuis, P josenkrantzi* and *H. glebosus* (from the Jurassic of Greenland-Harris 1931, 37), *H. wonnacotti, H. arcuatus, H. naiburensis* and *H. Hymenoptera* (From Jurassic of Yorkshire) have been described by Harris (1961). He has also described *H. amauros* from the triassic of Greenland (1939) and *H. solenotus* from the Triassic of England (1938).

The most completely known fossil representative of the Bryophytes is *Naiadita kinceolata* which has been descried by Harris (1938). The plant consisted of sparsely branched upright axis of 1 to 3 cm in height. The axis has no tissue differentiation but consisted of only parenchymatous cells. The axis usually at the base had considerable number of unicellular rhizoids with evenly thickened walls.

The axis had gemma cups; and each gemma cup had multicelluar gemma. Germination of gemma was apical. The leaves were arranged in 3/8 phyllotaxy. The lowermost leaves were linear, middle ieaves were lanceolate and topmost leaves were rounded. The leaves were single cell in thickness, without any midrib (Figure 5.53).

The archegonia were lateral, sessile and extra axillary and usually naked at first and afterwards archegonia were surrounded by a group of leaf like lobes. The archegonia had a single layered venter. There is no information about antheridia. The sporophyte was very small. It had small foot and large capsule. The jacket of the capsule was single-layered. There were lenticular spores. Columella and elaters were absent (Figure 5.54).

Naidita Lanceolata

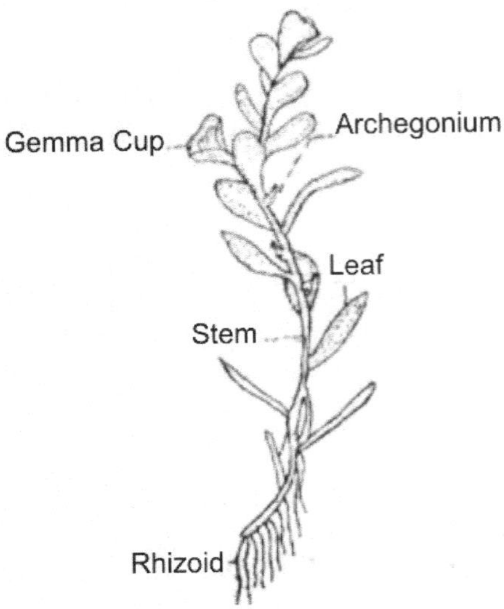

Figure 5.53

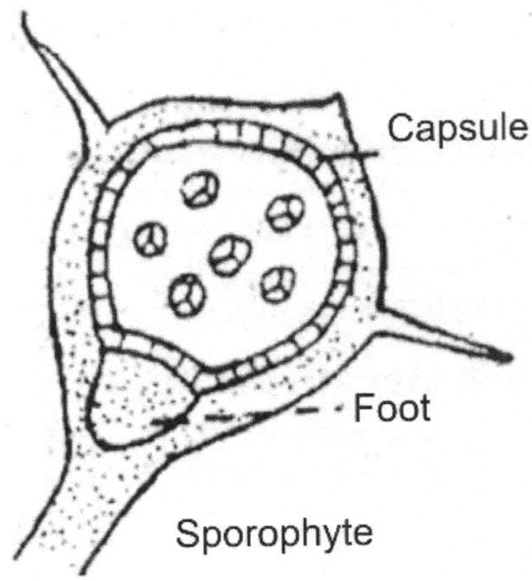

Figure 5.54

Harris was of the opinion that *Naiadita* resembles *Riella* in respect of the nature of rhizoids, and in the arrangement of archegonia and in the structure of the sporophyte.

There are also reports about the occurrence of a few plants like:

Ricciopsis sp. (resembling *Riccia*),

Marchantiolities sp. (resemblinig *Marchantia*),

Hepaticites cyathodoides (resembling *Cyathodium*),

Jungennannites (resembling Jungermanniales), from the Mesozoic era.

Fossil Bryopsida

Two species of Masses, namely *Muscites bertrandi* and *Musettes polytrichaceus* have been reported from Upper Carboniferous period. Neuberg (1956 and 1960) discovered, named and grouped them under Bryidae and Sphagnidae.

They are as follows:

Bryidae Sphagnidae

1. *Intia* sp. *Fungagia* sp.
2. *Salairia* sp. *Vorcutannularia* sp.
3. *Uskatia* sp. *Protosphagnum* sp.
4. *Polyssaieuia* sp.

5. *Bajdaiegia* sp.
6. *Buchtia* sp.

The geological history of Bryophyta only indicates that they are ancient group. The fossil forms of Bryophytes do not throw any light on the origin and evolution of the group. In this connection, Harris (1938) states that "we do not even find a suggestion whether they reached their climax of morphological development in the past, like so many Pteridophyte families.

Economic Importance of Bryophytes

The bryophytes, also called as liliputians amongst land plants, have shown their economic importance since stone age when pre historic man had made uses of feather moss (*Neckera crispa*). In modern age, these liliputians have attracted environmentalists, botanists, industrialists, horticulturists and pharmacists only because of their economic values.

The bryophytes include liverworts (Hepaticopsida) having over 300 genera and 6000-10,000 species, hornworts (Anthocerotopsida) having half a dozen genera and about fifty species and mosses (Bryopsida) having over 680 genera and about 15,000 species.

Economically the bryophytes can be grouped in two main catagories:

A. Direct use of bryophytes
B. Indirect use of bryophytes.

A. Direct Use of Bryophytes

1. Sphagnum and the Peat

The pieces of dead vegetable matter partially decomposed by water in the bog and gradually compressed and carbonized under pressure of the overlapping late deposits and water is called the *peat.* It is a brown or dark coloured spongy substance. *Sphagnum* and other mosses are the chief constituents of peat. Besides these peat contains remains of other plants growing in the bog such as the reeds, sedges, grasses, ferns and shrubs. Some peats however, consist mainly of the remains *of Sphagnum.* For this reason this moss is often called *peat moss.* These bog plants grow luxuriantly and in dense stands. Their basal older portions die. Thus large amounts of dead vegetable matter accumulate each year gradually filling the pond or lake. These deposits compressed and hardened by the weight of fresh deposits from above, in the course of time, attain considerable thickness. The compacted, partially decomposed and carbonized dead plant deposits are called the peat.

Uses of Peat

(*i*) Peat is used as a fuel. The thick deposits of peat are cut into blocks and dried. Being rich in carbon the dried peat blocks are used as fuel. In

Northern Europe, especially the Netherlands peat fuel is still used on a commercial scale. In the recent past the Europeans burned peat in the generation of power and there existed many factories in France for making illuminating gas from peat which has now been replaced by coal. Besides fuel, the following other uses of peat have been reported:

(ii) *Preparation of ethyl alcohol.* Odell and Hood (1926) reported that by chemical treatment cellulose in peat is broken down into sugar which through fermentation is converted into alcohol.

(iii) The Germans evolved a method to obtain ammonium sulphate as by-product in the production of gas from peat.

(iv) Nitrates, brown dye and tanning materials have been produced from peat.

(v) The other products of industrial use of peat are peat tar, ammonia and paraffin.

(vi) Davis (1911) reported that peat has been manufactured into paper, woven fabrics and artificial wood and employed as a mattress filler and as a bedding material for domesticated animals after cleaning the sticks and other coarse materials.

(vii) *Use in horticulture. Sphagnum* and peat are also of great use in horticulture. Peat is added to heavy (clayey) soils to improve their texture as it keeps them porous and prevents caking. When added to dry, sandy soils or other humus poor soils, it improves their water holding capacity and serves as a source of organic matter. Duffey (1954) observed that peat is the most important source of humus in the world. Dried sphagna are also used as garden mulch which helps to retain high soil acidity required by certain decorative and economically important plants. Cleansed peat and *Sphagnum* are extensively employed by horticulturists and florists as a substratum for seed germination, to start cuttings and to grow other plants and to propagate orchids. *Sphagnum* is also used as a packing for grafting scions to protect them against drying influence of the surrounding air.

(viii) Dried *Sphagnum* and moss peat for their ability to hold water tenaciously are remoistened and used as packing material for shipment of live plants, cut flowers, vegetables, perishable fruits, bulbs and tubers. Stokes (1931) observed that it (the packing material) retains enough moisture for a ling time and it protects against heat and cold. Peat is as well used as a packing material for fruits, fish, eggs and meat for cold storage.

(ix) Davis (1910) suggested that peat has a certain food value and thus has been used as a stock food in mixture with molasses.

(x) Davis (1946) reported that the various products of industrial use such as acetic acid, methyl alcohol, humic and allied acids, carbonic acid, paraffin, naphtha, montan wax and lignins used in making plastics have been obtained from peat by different methods.

(xi) Dacknowski-Stokes (1942) recommended peat moss as a suitable material for use in surgical dressings. Because of its great absorbent power and slight antiseptic properties carefully dried, cleansed and sterilized *Sphagnum* has long been employed for guaze to dress wounds and for making absorbent bandages in the treatment of boils and discharging wounds. Porter (1917) observed that the sphagnum dressings are cooler, softer and less irritating than those made with cotton.

2. Medicinal Use of Bryophytes

Information on medicinal use of bryophytes is scanty. Watt (1891) had reported to the medicinal use of *Marchantia polymorpha, Fagetella conica* and a few species of *Jungermannia, Anthoceros* and *Riccia.* Wren (1956) mentioned only *Polytrichum juniperinum* in this connection. Roig Y. Mesa (1945) mentioned that among liverworts *Marchantia polymorpha* has been used to cure pulmonary tuberculosis and afflictions of the liver. Hartwell (1971) reported that extracts of *M. polymorpha, M. stellata* and *Polytrichum commune* possess antitumour properties. However, the use of mosses in the treatment of ailments is more pronounced. Shiu-ying (1945) reported that dried *Sphagnum* is boiled in water. The decoction is used in the treatment of acute haemorrhage and diseases of the eye. According to Roque (1941) tea made of *Polytrichum commune* helps to dissolve stones of kidney and gall bladder. Peat tar, one of the byproducts of peat, has antiseptic properties and is used as a preservative. According to Grieve (1931), a distillate of peat tar known as *sphagnol has* been effectively used in the treatment of skin diseases. Hotson (1918 a) reported that Alaskan Indians prepared a healing ointment by mixing sphagnum leaves with grease or tallow. It was used in the treatment of wounds and cuts. The use of *Sphagnum* for making absorbent bandages in the treatment of boils and discharging wounds has been described above.

3. Antibiotic Activities of Bryophyta

There is scanty information about the occurrence of antibiotical substances in the bryophytes. However in recent years the antibiotical activities of bryophytes have drawn the attention of several workers. Hayes (1947) reported that aqueous extract of *Conocephalum conicum* is antibiotically active. Madsen and Pates (1952) and Pates and Madsen (1955) tested the antibiotic activities of four species namely *C. conicum, Dumortiera hirsuta, Sphagnum portoricense* and *S. strictum.* They found the first two to be active against *Candida albicans.* The two species of *Sphagnum* inhibited the growth of *Staphylococcus aureus* and *Pseudomonas aeruginosa.* Ramaut (1959) observed that the extracts of a species *of Sphagnum* inhibited the growth of *Sarcina lutea.* McCleary *et al.* (1960) reported *Anomodon rostratus* and *Orthotrichum rupestre* to be antibiotically active against *Micrococcus flavus, M. rubens, Streptococcus pyogenes* and *Candida albicans.* Later McCleary and Walkington (1966) tested the antibiotical activities of 50 species of mosses against two bacteria, *Gqffkeya tetragena* and *Staphylococcus aureus. Atrichium, Mnium,*

Polytrichum and *Sphagnum* showed pronounced antibiotic activity, 18 species were moderate to strongly active against one or both bacteria. The remainder showed slight but positive activity. Gupta and Singh (1971) reported that the petroleum ether extracts of two species of mosses namely *Barbula* and *Timella* are antibiotically active against 33 bacterial species which include gram negative, gram positive and acid fast bacteria. Later Bannerjee and Sen (1979) tested antibiotic activity of 52 species of bryophytes belonging to 40 genera against 12 micro organisms which included 3 gram positive (*Staphylococcus aureus, Sarcina lutea,* and *Bacillus subtills*), 5 gram negative (*Escherichia coli, Salmonella typhi, Vibrio Cholerae, Klebsiella pneumoniae* and *Pseudomonas aeruginosa*), one acid-fast bacterium (*Mycobacterium phlei*) and 3 fungi (*Curvularia lunata, Aspergillus niger* and *Heminthosporium oryzae*). They came to the conclusion that a variety of antibiotic substances occur among bryophytes.

4.

During the last decade several unsaturated lipids, fatty acid esters, flavonoids, triterpenoids, phenolics and other chemical substances have been reported to be present in the bryophytes.

5. Use of Bryophytes in Experimental Botany

The liverworts and mosses have palyed an important role as research tools in the various phases of botany such as genetics, experimental morphology and physiology. The mechanism of sex determination in plants was discovered for the first time in a liverwort *Sphaerocarpos.* Allen's studies (1935, 1945) on the genetics of bryophytes and Wettstein's investigations (1932) on hetroploid bryophytes are classics. Berie (1964) made experimental studies on polyploidy in liverworts. Important contributions on desiccation and regeneration, which are their chief attributes, have been made in the recent past by various investigators. Giminghan and Birse (1957) studied correlation between growth form and habitat in bryophytes and Birse (1957) made experimental studies on growth forms in mosses.

6. Bryophytes as a Source of Food

Bryophytes are not directly used as human food. However, Landley (1856) made mention of *Sphagnum* as a wretched food in barbarous countries. Read (1946) listed peat moss as a famine food in China. Haines (1877) recorded Laplanders having made use of *Sphagnum* as an ingredient in the preparation of bread. There are, however, examples of animals useful to man which use bryophytes as a food source. Lid and Miedell (1933) reported that moss capsules such as that of *Bryum* and *Polytrichum* constitute the chief diet of the Norwegian grouse chicks. The capsules of moor-inhabiting mosses are the favourite food of the red grouse chicks. Collinge (1927) recorded that some English birds such as the field fare, song thrush and black bird include some mosses as a small part of their regular diet. Hone (1934), Hawden and Palmer (1922) observed that Alaskan reindeer grazes upon *Polytrichum* sp., *Aulacominum turgidum,* and *Hylocomium alaskanum* besides lichens and grasses in summer. In winter they include *Sphagnum fimbriatum* and *Dicranum* spp. in addition to the above.

7. Bryophytes as Pollution Indicators

Rubling and Tyler (1979) and Ferguson *et al.* (1978) showed that both alive and air dried mosses particularly *Sphagnum* can absorb metals. The accumulation of heavy metal cations in mosses enables them to be used as atmospheric and aquatic pollution indicators and in minrological surveys.

8. Miscellaneous Uses of Bryophytes

Dicranum elongate was used by Cree Indians for lampwicks (Chamberlain, 1901) and the Eskimos of N.E. Labrador used *Rhacomitrium lanuginosum* for the same purpose (Hantzsch, 1931). Evans and Nichlos (1908) reported that foliose liverworts have been employed as a packing material for living plants in the tropics and some masses for vegetables. Mosses have been as well used as a decorative material for pillows. Brail and Moyle (1976) reported that birds employ fragments of gametophores of mosses and liverworts in the construction of their nests.

B. Indirect Use of Bryophytes

1. As Aids in Soil Conservation

The mosses prevent sheet erosion of soil. They grow in dense stands forming a mat or carpet-like structure. This thick moss carpet serves a twofold function. Firstly it bears the impact of the falling rain drops and secondly it holds much of the water that falls. Consequently the amount of run-off is considerably reduced. Besides there is no or little turbidity in the run-off. Conard (1935) demonstrated that the turbid water flowing through the moss mat is shorn much of its sediment. The inter-twined moss stems and the underground rhizoids bind the soil particles together to a considerable depth (6-8 inches) so firmly that there is no or little erosion even on a steepy hill side. Grout (1912) held that even the moss protonema mat checks soil erosion.

2. Formation of Soil and Development of Vegetation Cover

The lichens and mosses play an important role in soil formation. Both are slow but efficient soil formers. The lichens, however, are the pioneers to colonise barren, bare rocky surfaces where no other plants can grow. The lichen thalli secrete organic acids which gradually dissolve and disintegrate the rocks to which they cling. The rock particle together with the dead and decaying older parts of the lichen thalli form fertile soil. Mosses make their appearance when sufficient amount of this fertile soil gathers in the crevices of the rock surface. The progressive growth and death of the older parts of the mosses adds to the substratum begun by the lichens. Dust and debris blown by the wind rapidly accumulate between the erect moss stems. Weaver and Clements (1938) reported that in the course of time, the depth of soil under the moss mats may be more than one inch. The moss mat collects sufficient moisture and contains humus to form a suitable substratum for the growth of many rock loving herbaceous plants. In fact Schroeter (1926) recorded that *Saxifraga cotyledon* germinates only among the mosses. The other rock loving species which usually appear in the moss mats are *Sedum, Thymus, Saxifraga aizoon* and others. To sum up in the development of vegetation cover over rocky surfaces, the mosses play a vital role. The moss stage follows the lichen stage and precedes the herbs. The latter, in turn, may be succeeded by shrubs and trees.

3. Bog Succession

Weaver and Clements (1938) remarked that mosses play an important role in bog succession from open water to climax forest. The mosses especially the peat mosses established on the banks of lakes and other shallow bodies of water extend inwards and grow over the surface of water with their stems intertwined to form thick mats. These surface mats over bodies of water give the appearance* of solid soil. Such areas are called *quacking bogs.* The thick moss mat because of the moisture and humus forms a suitable substratum for the germination of seeds of various species of hydrophytic plants. The older parts of these plants gradually die and settle to the bottom. In the course of time, these lakes and shallow bodies of water become filled up with partially decomposed old parts of mosses and other hydrophytic plants. Thus the areas which originally were sterile sheets of water, become converted into solid soil supporting vegetation. The mosses and hydrophytic angiosperms eventually disappear. They are replaced by forest growth of mesophytic type. The mosses thus play a vital role in changing the landscape.

4. Role as Rock Builders

Certain mosses (*Bryum, Hypnum, Fissidens* and others) growing in association with other aquatic plants (such as *Chara* and many other algae) play a remarkable role as rock builders. These plants grow in shallow waters of lakes, streams and springs which contain a large amount of calcium bicarbonate. The plants bring about decomposition of bicarbonic ions by abstracting free carbon dioxide. The insoluble calcium carbonate precipitates. This insoluble mineral, on exposure, hardens forming calcareous (lime) rock-like deposits around these plants. These travertine deposits continue to grow by the aid of mosses and algae growing in the water extending over areas of several hundreds square feet. The travertine rock deposits are extensively used as a building stone.

Though large number of species of Bryophytes are available, only a few plants have been reported to have economic value. They are: *Sphagnum, Polytrichum* and *Marchantia.* Generally the liverworts, Mosses and Lichens are supposed to be the pioneers in establishing vegetation where other vegetation seems to be practically impossible. They are considered as rock builders. They colonize the barren rocks and exposed areas of hills and increase humus content of the soil and make the soil suitable for growth of other plants and for cultivation. They form an extensive carpets to hold the soil and prevent soil erosion. '

The dead and decomposed, compacted and carbonized deposition of *Sphagnum* plants are called peat. Dried peat blocks are used as fuel. Peat is used in generating power and for making illuminating gas and Ammonium Sulphate is obtained as a biproduct.

By chemical treatment of peat cellulose, Ethyl Alcohol is obtained. Nitrates, brown dye and tannin materials have been produced from peat. The woven fabrics and artificial wood obtained from peat are used as mattress filler and bedding material for domestic animals.

- ☆ Peat when added to soil, it improves soil texture, water holding capacity and serves as a source of organic matter. Most of the Mosses, especially the dried peat Mosses are used as packing material to pack bulbs, cuttings, seedlings for shipment.
- ☆ As the peat Mosses have the power to absorb and hold water, they are extensively used in seed beds and green houses.
- ☆ The *Sphagnum* species are slightly antiseptic and possess superior absorptive power. On account of these properties they may be used for filling absorbent bandages in the place of cotton in the hospital.
- ☆ Recently antibiotic activities have been reported from the following plants. Some species of *Conocephalum, Dumortiera, Sphagnum* and *Polytrichum* showed pronounced antibacterial activities.
- ☆ Some of the liverworts and Mosses have been used as experimental plants. The mechanism of sex determination in plants was discovered in *Sphaerocarpos* for the first time.
- ☆ *Sphagnum* species are used as atmospheric and aquatic pollution indicators.

The peat Moss is used as reched food in uncivilized countries and famine food in China.

Points to Remember

1. The bryophytes are an amphibious nature. It has two generations one is gametophyte and another is sporophyte. The plant body are commonly called gametophytes. Based on the structure of the plant body is classified into liverworts, hornworts and Mosses.
2. The plant body is not differentiated into root, stem and leaf. Because it is thalloid in nature. Dorsiventral flattened, dichotomously branched, air pore is present in upperside, rhizoids are present in lowerside.
3. Rhizoids are two types namely, smooth walled and pegged rhizoids. In internal structure it shows two regions one is storage region and another is photosynthetic region covered by epidermis in thalloid liverworts.

4. The thallus structure of the leafy liverworts internal simplification and external complication. In the case of Mosses the plant body some What differentiated into root like, stem like and leaf like structures.
5. After full maturity reproduction by three methods vegetative, asexual and sexual reproduction.
6. The gametophyte produced the male and female gametes during the reproduction. The male gametes are called antheridium and female gametes are called archegonium. The antheridial cells are differentiated into outer cells, basal cells and antherozoids.
7. The archegonium cell consist of venter canal cell, neck canal cell and egg cells. After fertilization zygote is formed.
8. The zygote is the first cell of sporophyte. The sporophyte is differentiated into foot, seta and capsule. In the lower forms the capsule will become reduced and the sporophyte somewhat physiologically independent in Stonworts and Mosses.
9. The sporophyte produce spores it germinates to produce protonema and develop into a gametophyte. Hence, it showed two phases namely gametophytic pahse (n-haploid) and sporophytic phase (2n-diploid) and altenate with each other.
10. These process is called alternation of generation. The Bryophytes are posses heteromophic alternation of generation.

SELF LEARNING MATERIALS

1. Introduction

- ✰ Morphology of Riccia plant body is a small green thalus, its dorsiventral and grown prostrate on the ground.
- ✰ It branches freely by dichotomy and frequently takes on a circular form.
- ✰ Each thallus lobe has a distinct midrib, a dorsal media groove or furrow and ends in a terminal notch the areole and the gemma cups are absent.
- ✰ Morphology of Marchantia is dichotomously branches thallus fleshy in texture. Each thallus lobe has a distinct midrib but no dorsal furrow, it ends in a terminal notch.
- ✰ Presence of gemma cups is an other distinguishing features.
- ✰ Porella – The plant body is generally epiphytic in habit. The thallus is prostrate and dorsiventral.
- ✰ It consist of a more or less branched central axis with leaves arranged in three rows.

- These are two rows of lateral and symmetric leaves on the upper surface of the axis. The third row of reduced leaves is on the lower side of the axis.
- The dorsal leaves are bilobed and incubously arranged. The branching of axis is monopodial.
- In Anthoceros: Plant body is simple some what fleshy dorsiventral, branching is dichotomous but due to unequal growth of the two parts of dichotomy thallus usually becomes irregular in outline.
- The thallus is usually inhabitated by Nostoc. Colonies seen as small dark, blue green speck or sports in surface view.
- Sphagnum grow in very wet places as semi aquatics and also submerged confined to acidic water-logged habitats.
- The leafy gametophyte is a perennial plant. It has an upright stem. Individually the erect stem is weak and fragile.
- The plant is erect and radial in organization. There is no distinction into upper and lower surface. The mature gametophytes lack rhizoids.
- In Funaria – it grows in dense patches in moist situation such as damp soil, on the wall, on moist rocks and on lands burnt by fires.
- The plant body is leafy, green upright and radial in symmetry. The leaves are spirally arranged. They are sessile oblong-ovate with a smooth margin pointed apex and distinct midrib. The branching is monopodial.
- Internally stem shows certain amount of tissue differentiation. The cells are arranged in three distinct regions: epidermis, cortex and central cylinder.
- Epidermal cells are thick walled and lack chloroplasts. There are neither any air pores nor stomata.
- Cortex: It is several cell layers thick and consist of large, thin walled cells. They are compactly arranged and in your stem contain chloroplast.
- The nature portion of the stem have the cortex differentiated into an outer thick – walled cortex and an inner thin-walled cortex. There are no air spaces.

Central Cylinder

- It occupies the core or the stem and consist of vertically elongated thin walled cells. They are narrow and compactly arranged.
- It provides mechanical support to a certain extent and functions in conduction.
- The so-called leaf consists of a single layer of cells rich in chloroplasts.

The midrib in several cells thick, it contains a narrow strand of thin walled, elongated cells.

2. Objective

- ☆ In porella the plant body is a leafy thallus which is an elaborate, delicate structure. The appearance of leaf-like expansions arising from the central axis.
- ☆ The Slender, central axis or the stem is branched in a bi- or tri-pimmate manner. The branching is always monopodial dichotomous branching which is so characteristic of other liverworts is absent in the Jungermanniales.
- ☆ Simple, smooth-walled rhizoids spring from the ventralside of the axis near the bases of ventral leaves and fix the plant to the substratum. They are unicellular.
- ☆ Apical growth takes place by means of a single, large apical cell pyramidal in shape.
- ☆ The antheridia occur singly in the axis of leaves on the antheridal branches which spring approximately at right angles from the central axis and bear closely imbricated light green leaves which are often called the perigonial leaves or bracts.
- ☆ The archegonia occur in a cluster at the apex of a very short archegonial branch. The latter bears only two to four perichaetial leaves.

3. Content

- ☆ Bryophyte, vegetative reproduction, malegametophytic plant, female gametophytic plant.
- ☆ Antheridium – Archegonium – Antherozoid Egg
- ☆ Fertilization – Zygote – Embryo, Sporangium.
- ☆ Spore mother cell – Meiosis – Spores – Protonema.

4. Revision Points

- ☆ **Fossil Bryophytes:** Plant body is simple and small in size. Plants are delicate and gametophytic. The surface of thallus or leaf is not cuticularised.
- ☆ **Origin of Bryophyta:**
  - (i) Evidence from comparative morphology of the living plants.
  - (ii) Evidence from ontogeny of the living plants.
  - (iii) Evidence based on analogies with the living plants of other groups.

☆ **Systematic Position**

(i) Thallus simple, green, thalloid plant body.

(ii) Lack of Vascular tissue and absence of roots.

(iii) Predominant plant in the life cycle being the gametophyte.

(iv) Autotrophic mode of nutrition.

☆ **Biological Importance:** Origin of land habit among plants

(i) Compact plant bodies usually protected by epidermis containing air pores.

(ii) Primitive type of conducting strand.

(iii) To facilitate absoption of water and salts from the soil the plant body is provided with rhizoids.

(iv) Jaucketed, multicellular sex organs.

(v) Thick walled, wind disseminated spores.

☆ **Distribution of Bryophytes:** Pande (1958) divided the Indian Subcontinent into four major Bryogeographical units which are as follows:

1) The Himalayan Regions: Western Himalayan Territory - It extends from the Western boundry of Nepal to Kashmir.

Eastern Himalaya – Eastern parts of Mountains of Indian territory separating from India from neighboring countries like Burma, Bangaladesh and China.

2) Punjab and the West Rajesthan: Central India and the Gangetic Plains.

3) Southern Zone: It divides into two pars West Coast Region, it represent total area of Western Ghats. East Coast Region and Decan Plateu: It represent Eastern Ghats, Nilgiris.

☆ **Physiology of Bryophytes:**

1) Water Relations: Water absorption takes place either from the plant surface or through the rhizoids Bryophytes have been grouped into three main categories on the basis of their water relations.

a) Ectohydric: Which are capable of absorbing water or solutes by surface absorption only such sps. do not possess a well developed conducting.

b) Endohydric: Which are able to strands absorb water or solutes mainly through their Rhizoids which are internally conducted through well defined conducting strands.

c) Mixohydric: Species in which water and solute re absorbed by both the ways that is by ectohydric and endohydric methods.

- ☆ **Oil bodies:** The oil-bodies are distinctive and characteristic features of only liverworts. They are absent in hornworts and mosses.
- ☆ Most interesting feature is that such type of oil-bodies are not found elsewhere in the whole of Plant Kingdom.
  - i. Oil-bodies protect the plant body from bright light.
  - ii. They provide mechanism to protect the plant against dessication.
  - iii. Oil-bodies help the plant to survive under extreme cold environments. They are also known as "Cold Protection Bodies".
  - iv. Presence of oil bodies – Liverworts are not attacked by snails and insects.
- ☆ **Apogamy in Bryophytes:** Apogamy means the development of a Sporophyte (embryo) out of any gamatophytic cell without the union of male and female gametes.
- ☆ **Apospory in Bryophytes:** Apospory means the development of a gametophyte from any part of the sporophyte without any reduction division.

5. Intext Question

1. Write a comparative account of the organization of sporophyte in Riccia. Anthoceros and Funaria.
2. Explain the method of fertilization in moss and describe the post-fertilization changes in moss plant with the help of suitable diagram.
3. Enumerate significant steps in the life cycle of Funaria.
4. Write short notes on: (i) Gametophyte of Moss, (ii) T.S. of Funaria stem, -(iii) V.S. leaf of Funaria, (iv) Spore dispersal in Funaria, (v) Protonema.

6. Summary

- ☆ Amphibians *i.e.,* grow in moist shady places as well as in water.
- ☆ The thallus, excepting in the protenema stage, is never filamentous, its made up of sheets of parenchyma cells.
- ☆ Sexual reproduction is Oogamous only.
- ☆ Sex organs covered by a sterile jacket.
- ☆ The female sex organs is the flask-shaped Archegonium which is characteristic of land plants only.
- ☆ Embryo produced from zygote, the zygote is neither libeated nor passes into the resting stage.
- ☆ Sporophyte, developed and differentiated into foot, seta, and capsule.

☆ The Sporophyte is organically attacked to and is nutritionally dependent upon the gametophyte.

7. Terminal Exercise

1. Which of the following is not found in Marchantia

   a) Perigynium b) Perichaetium

   c) Peritoneum d) Paraphysis

2. Barrel shaped pores are found in

   a) Riccia b) Marchantia

   c) Pellia d) Anthoceros

3. In Marchantia the chloroplast is

   a) Cup shaped b) Disc shaped

   c) Star shaped d) Ribbon shaped

4. In which of the following tuberculate rhizoids are present.

   a) Funaria b) Anthoceros

   c) Marchantia d) None of the above

5. Elaters of_______ help in dispersal mechanism.

   a) Riccia b) Selaginella

   c) Funaria d) Marchantia

8. Assignments

☆ Give a brief account of phyletic evolution of Jungermanniales.

☆ Give an illustrated account of leaves in Jungermanniales.

☆ Write a brief account of the role of protonema in the life history of moss plant?

☆ With the help of diagrams write the characteristics of rhizoids in Liverworts and mosses.

☆ Discuss: which genus has largest spore among hepacticpsida and bryopsida.

9. Learning Activities

☆ Give an account of the vegetative propagation and modes of perennation in Bryophyta.

☆ Discuss the Evolutionary Tendencies in the gametophytic and sporophytic generation of Bryophytes.

- ☆ Give an account of Fossil Bryophytes. Write a short account on the origin of Bryophyta studied by you.
- ☆ Explain condensation theory of the evolution of gametophyte in Bryophyta.
- ☆ Having the origin of land habitats among plants. Justify the Bryophytes as amphibians in plant kingdom.

10. Key words

- ☆ Vegetative propagation
- ☆ Gamatophytic generation – Sporophytic generation
- ☆ Condensation theory – Archesporium – Sporogenous tissue
- ☆ Fossil Bryophytes, Apogamy, Apospory
- ☆ Bryophytes as Ecological Indicators.

Model Questions

Section – A

Each answer should not exceed 50 words

1. Antherdiophore
2. Archegoniophore
3. Gemmae
4. Elaterophore
5. Elaters
6. Pseudoelaters
7. Riccardia
8. Archegonia
9. Antheridia
10. Protonema
11. Operculum
12. Peristome
13. Columella
14. Calyptra
15. Bog moss (or) Peat moss
16. True moss

17. Heteromorphic alternation of generation
18. Involucre
19. Incubous arrangement
20. Succubous arrangement
21. Amphibians
22. Naiadita lanceolata
23. Carpocephalum
24. Amphigastriya
25. Sporophyte

Section – B

Each answer should not exceed 300 words

1. Write an account of the fossil Bryophytes.
2. Give an account of the asexual reproductive stucture of the Bryophytes.
3. Discuss the evolutionof gametophyte in thalloid and leafy liver worts.
4. How are Bryophytes classified?
5. Discuss briefly the conducting system in Bryophytes?
6. Describe the sporophyte of Anthoceros and comment on its salient features.
7. Give a brief account of Economic importance of Bryophytes.
8. Discuss the origin and evolution of Bryophytes.
9. Give an account of the sporophyte of Mosses.
10. Give an account of vegetative reproduction in Bryophytes.
11. Give an account of the sexual reproduction in the Bryophytes.
12. Discuss the evolution of gametophyte in thalloid and leafy liverworts.

SECTION-C

Each answer should not exceed 1,200 words

1. Write an essay on the progressive and retrogressive evolution of Bryophytes.
2. Write an essay comparing the gametophytes of the liverworts with that of the Mosses.
3. Trace the origin and evolution of sporophytes in Bryophytes.

4. Describe the diagnostic features of the following forms.
   a) Marchantia
   b) Pallavicinia
   c) Riccardia
   d) Bryum
5. Give an account of spore dispersal mechanism in Bryophytes.
6. Discuss on fossil Bryophytes.

Selected Bibliography

1. Sharma, O.P. (2015). Text book of Algae, Tata Mc Graw – Hill Publications Pvt Ltd., New Delhi.
2. Sureshkumar and Khaling Mikawfrawng (2014). Text book of Bryophyta, Sonali Publications, New Delhi.
3. Vau Deu Hock, C., D.G. Mann and H.M.Johns (2019). Algae – An Introduction to Phycology, Cambridge University Press, New Delhi.
4. Vashishta, B.R., A.K. Sinha and Adarsh Kumar (2016). Botany for Degree Students – Bryophyta. S. Chand and Company Ltd., New Delhi.
5. Vashishta, B.R., A.K. Sinha and Adarsh Kumar (2015). Botany for Degree students – Algae. S. Chand and Company Ltd., New Delhi.

Reference Books/Supplementary Reading

1. Fritsch, F.E. (1972). Structure and Reproduction of Algae I and II, Cambridge University Press.
2. Rashid, A. (2007). An Introduction to Bryophyta – Vikas Publications, New Delhi.
3. Watson, E.V. (2015). The Structure and Life of Bryophtes. United Book Prints, New Delhi.
4. Vashishta, B.R., A.K. Sinha and V.P. Singh (2014). Botany for Degree Students – Algae, S. Chand and Company Pvt. Ltd., New Delhi.
5. Sivakumar, K. (2016). Algae - A Practical Approach, MJP Publishers, Triplicane, Chennai

Index

D

E

R

S

9 789359 195261

Printed at Repro India Ltd.